Biometrics

Routledge Studies in Science, Technology and Society

1. **Science and the Media**
   Alternative Routes in Scientific Communication
   Massimiano Bucchi

2. **Animals, Disease and Human Society**
   Human-Animal Relations and the Rise of Veterinary Medicine
   Joanna Swabe

3. **Transnational Environmental Policy**
   The Ozone Layer
   Reiner Grundmann

4. **Biology and Political Science**
   Robert H. Blank and Samuel M. Hines, Jr.

5. **Technoculture and Critical Theory**
   In the Service of the Machine?
   Simon Cooper

6. **Biomedicine as Culture**
   Instrumental Practices, Technoscientific Knowledge, and New Modes of Life
   Edited by Regula Valérie Burri and Joseph Dumit

7. **Journalism, Science and Society**
   Science Communication between News and Public Relations
   Edited by Martin W. Bauer and Massimiano Bucchi

8. **Science Images and Popular Images of Science**
   Edited by Bernd Hüppauf and Peter Weingart

9. **Wind Power and Power Politics**
   International Perspectives
   Edited by Peter Strachan, David Lal and David Toke

10. **Global Public Health Vigilance**
    Creating a World on Alert
    Lorna Weir and Eric Mykhalovskiy

11. **Rethinking Disability**
    Bodies, Senses, and Things
    Michael Schillmeier

12. **Biometrics**
    Bodies, Technologies, Biopolitics
    Joseph Pugliese

Biometrics

Bodies, Technologies, Biopolitics

Joseph Pugliese

Routledge
Taylor & Francis Group
New York London

First published 2010
by Routledge
605 Third Avenue, New York, NY 10017

Simultaneously published in the UK
by Routledge
2 Park Square, Milton Park, Abingdon, Oxon OX14 4RN

Routledge is an imprint of the Taylor & Francis Group, an informa business

First issued in paperback 2012

Typeset in Sabon by Taylor & Francis

Library of Congress Cataloging-in-Publication Data
Biometrics : bodies, technologies, biopolitics / edited by Joseph Pugliese.
p. cm. – (Routledge studies in science, technology, and society; 12)
Includes bibliographical references and index.
1. Biometric identification. I. Pugliese, Joseph, 1959–
TK7882.B56P84 2010
006.4–dc22 2009051431

ISBN 13: 978-0-415-81104-0 (pbk)
ISBN 13: 978-0-415-87487-8 (hbk)
ISBN 13: 978-0-203-84941-5 (ebk)

For Sebastian

Contents

List of Figures

Acknowledgments

This book emerged from work that began, long ago, with a Nietzschean analysis of dermographic inscriptions on the body. Magnetised by the indissociable relation between the body and cultural inscription, I proceeded to track the cultural politics of the body under the power of assimilationist regimes of race, identity and nation. This, in turn, led me to investigate the intersection of science, law and race in the context of genetics and forensic pathology. It was from that juncture that biometrics beckoned.

A number of people have been crucial in helping me bring this book to fruition. I am especially grateful to Suvendrini Perera, who encouraged me to pursue this project from her very first reading of my work on biometrics; our dialogues and exchanges have been a constant source of affirmation; her research has been a critical point of reference. My thanks to Benjamin Holtzman at Routledge for his enthusiastic response to my proposal and to the anonymous referees who underscored his belief in the project. I'm grateful to Penny Pether, David Caudill and Peter Goodrich for their generous support and for encouraging and enabling my repeated transgressions in the field of law. Judith Grbich has played a crucial role in the field of Australian intellectual life; her journal, *The Australian Feminist Law Journal*, has been at the forefront in publishing innovative and risk-taking political work that would otherwise not see the light of day. My thanks to Judith for publishing my first paper on biometrics and for her continued support of my work. Susan Stryker's coining of the term "somatechnics" proved invaluable in clarifying my approach to bodies and technologies—I'm grateful to Susan and the members of the Department of Critical and Cultural Studies at Macquarie University who contributed to the theorising of this term.

I wish to express my thanks to Maria Giannacopoulos and Lara Palombo for their practical assistance and unwavering political commitment, as well as acknowledging Anthony Burke, Anne Cranny-Francis, Cathy Hawkins, Ivor Indyk, Goldie Osuri and Juliet Rogers for their various modes of support. In addition, I express sincere thanks to Paul Collier, Victor Lee, Nicholas Orlans and Charlie Wilson for generously sharing their expert knowledge of biometrics.

Iole and Carmelo have been indefatigable in their support, even through the worst of times; their unqualified generosity has nourished me throughout the years. I'm grateful also to Raffaella, Larry and John who offered support during difficult times. In the midst of everything, Sebastian's impassioned engagement with issues of social justice has been an inspiration. Trish has always been there for me: steadfast, caring and always enabling in the face of trauma and turmoil. This book would not have been possible without her.

The two following translations are by the author: Marcello Levi Bianchini. 1906. "La mentalità della razza calabrese: saggio di psicologia etnica," *Rivista di Psicologia Applicata alla Pedagogia ed alla Psicopatologia* 2: 14–18; and Cesare Lombroso. 1889. *L'Uomo Delinquente in Rapporto all'Antropologia, alla Giurisprudenza ed alle Discipline Carcerarie*, 2 volumes. Torino: Fratelli Bocca.

A number of sections in this book were previously published in journals and have been reproduced with permission. They include: "*In Silico* Race and the Heteronomy of Biometric Proxies: Biometrics in the Context of Civilian Life, Border Security and Counter-Terrorism Laws," *The Australian Feminist Law Journal*, 23 (2005): 1–34; "'Demonstrative Evidence': A Genealogy of the Racial Iconography of Forensic Art and Illustration," *Law and Critique*, 15.3 (Springer, 2005): 287–320; "Biometrics, Infrastructural Whiteness, and the Racialized Zero Degree of Non-Representation," *boundary 2*, 34.2 (2007): 105–33 (Copyright 2007, Duke University Press. All rights reserved. Reprinted by permission of the publisher.); "Biotypologies of Terrorism," *Cultural Studies Review*, 14.2 (2008): 49–66; "Preincident Indices of Terrorism: Facecrime and Project Hostile Intent," *Griffith Law Review* (2009): 314–30 (published with permission from *Griffith Law Review* and Griffith University); "The Alleged Liveness of Live: Legal Visuality, Biometric Liveness Testing and the Metaphysics of Presence," in Anne Wagner and Richard K. Sherwin (eds.), *Treatise on Legal Visual Semiotics* (Springer, 2010).

The images from Cesare Lombroso's *L'Uomo Deliquente* are reproduced courtesy of the University of Otago Library.

Introduction: Biopolitics Of Biometrics

In 2001, biometrics was named by the influential *MIT Technology Review* "one of the 'top ten emerging technologies that will change the world'" (Woodward et al. 2003, xxiii). Soon after the September 11, 2001 attacks on New York and Washington, D.C., the biometrics industry was further catapulted into the international spotlight by governments eager to deploy biometric systems in order to allay concerns about national security. Situated in this heady context, biometrics is too often represented as a technology with extraordinary powers and, simultaneously, as a technology that has only recently emerged on the historical stage. As I proceed to demonstrate in Chapter 1, biometrics has a long and complex historical genealogy that must be tracked back to the nineteenth century and the concomitant emergence of *biopolitics* (a term I will presently discuss in some detail). Once located within this complex genealogy, the extraordinary powers attributed to biometric technologies, in particular, ongoing pronouncements from both its manufacturers and advocates that they are neutral and objective technologies unmarked by issues of prejudice or bias, are shown to be untenable.

In the course of this book, I proceed to argue that biometrics, as a technology of authentication and verification, achieves its signifying status only by being situated within relations of power and disciplinary techniques predicated on individuating, identifying, classifying and distributing the templates of biometrically enrolled subjects across complex political, social and legal networks. As such, biometrics is a technology firmly enmeshed within relations of *biopower*. The emergence of the biopolitical state, Michel Foucault (2008, 34) underscores, is marked by the installing of the veridictional question, "the question of truth," at the core of its criminological operations. The question: "What have you done?" is now replaced with the question: "Who are you?" (Foucault 2008, 34). The foundational question of biometric technologies is precisely "Who are you?" The key premise of this book is that this biometric question is repeatedly made coextensive, in the biopolitical operations of biometric technologies, with *what you are*. As I discuss in some detail in the chapters that follow, the answer to the question "Who are you?" pivots on the specificity of a subject's embodiment and her or his geopolitical status.

What you are—a person of colour and/or an asylum seeker—determines the answering of who you are.

In marking the indissociable relation between this foundational biometric question and the exercise of biopolitical power, I argue that, as contemporary instrumentalities of biopolitics, biometric technologies are inscribed with infrastructural relations of disciplinary power underpinned by normative categories of race, gender, (dis)ability, sexuality, class and age. As I contend throughout the course of this book, it is the invisibilised, because infrastructural, status of these normative categories that allows for ongoing pronouncements about the technology's neutral and non-discriminatory capability. Following from this premise, my concern is to begin to map the ways in which biometric technologies operate as systems for the discrimination of non-normative subjects, including people of colour, refugees and asylum seekers, transgender subjects, labourers and people with disabilities. Framing biometrics within the conceptual schema of biopolitics will enable the fleshing out of the complex intersection of bodies, subjects, technologies and power and the consequent articulation of the lived effects of biometrics as apparatuses of biopower.

In what follows, I proceed to delineate the major politico-theoretical concerns of this book and to clarify the key terms that constitute its critical apparatus of inquiry.

Biometrics as Evidentiary Technologies

Biometric systems are technologies that scan a subject's physiological, chemical or behavioural characteristics in order to verify or authenticate their identity. There are three key steps in the operation of a biometric system. In the first instance, a subject presents herself to a biometric system in order to "enrol" in the system. This moment, which in the language of biometrics is called "presentation," allows the biometric imaging system to capture an image of the particular "imprint" of the subject, for example, in facial-scan, an image of their facial characteristics. This image is then digitally converted through the use of particular algorithms into what is called a "template". The template of a subject's unique biometric characteristics is then stored in the system's server. Every time a subject attempts to gain access to either a physical site, such as a secure office space, or access restricted information stored in a server, the subject must present themselves before the biometric system, which re-scans their facial details and proceeds to compare them with the original scan (or scans) that has been algorithmically processed and stored in the server as the reference template. Whether or not the new scan matches the template determines whether or not the subject is allowed access to the restricted information or secure area.

Biometrics can be succinctly characterised as a technology of capture: that is, the technology is fundamentally predicated on capturing images of subjects. It is this process of visual capture that enables the processes of template creation and consequent verification or authentication of a subject's identity. The success or failure of a biometric system is predicated on the assumption that a subject can provide a biometric authenticator (for example, a fingerprint) that is unique to that subject and that cannot be replicated by an impostor (as I discuss in some detail in Chapter 4, this assumption is, for many reasons, flawed). In the process of attempting to clear the security procedures of a biometric system, a subject, who has previously established a biometric template within the system, provides a biometric authenticator that is, in turn, verified by the system's verification procedure; this entails an "algorithm that compares an authenticator with a verifier" (Woodward et al. 2003, 4–5). The key terms that frame biometric discourse—"authentication" and "verification"—underline the manner in which biometric technologies transmute a subject's corporeal or behavioural attributes into evidentiary data inscribed within regimes of truth.

Regimes of Truth

Biometrics' conceptual and technological conditions of emergence can be traced back to the nineteenth century. As I discuss in Chapter 1, biometric technologies, in all of their proto-manifestations, were formulated and implemented in the context of the emergence of a new form of politics: *biopolitics*. Driving the development of biometrics in the nineteenth century was a shift in the type of question posed by the state to its target subjects. As noted earlier, the emergence of the biopolitical state is marked by the installing of the veridictional question, "the question of truth," at the core of its criminological operations. As biometric technologies are fundamentally concerned with questions of verification and authentication, they must be seen as technologies of "truth." They convert physiological (for example, an iris scan) or behavioural (for example, a subject's gait signature) information into evidentiary text that will proceed to disclose the "truth" of a subject's identity, the "truth" of a subject's authenticity and, even, the "truth" of their intent (see Chapter 3). Moreover, the category of truth is, in biometrics, seen as what must be captured by the technology; so, for example, the effectiveness of a particular biometric technology is gauged by its ability to "preserve" what biometricians term the "ground truth" of a subject's physiological or behavioural attribute (Parziale and Chen 2009, 107). As such, biometric systems are inscribed as evidentiary technologies productive of "truth."

I have been placing the term "truth" in quotation marks in order to bring into focus its status as a category that is not simply self-identical but, rather, is complexly produced by relations of power and knowledge. Drawing on the work of Foucault (1980, 131), truth emerges as a category that achieves its "truth status," so to speak, through the intertwining of regimes of power and knowledge that discursively determine and delimit the truth of a particular subject:

> Truth is a thing of this world: it is produced only by virtue of multiple forms of constraint. And it induces regular effects of power. Each society has its régime of truth, its 'general politics' of truth: that is, the types of discourse which it accepts and makes function as true; the mechanisms and instances which enable one to distinguish true and false statements, the means by which each is sanctioned; the techniques and procedures accorded value in the acquisition of truth; the status of those who are charged with saying what counts as true.

So, for example, as technologies of truth, biometric systems are underpinned by the authorising discourse of science. The discourse of science is enabled to make its truth claims because of its deployment of the scientific method. The scientific method is based on empirical and observable evidence that is gained through formalised practices of experimentation that can be replicated by other scientists in order to verify the claims made by a particular scientist. Furthermore, when viewed in Foucauldian (1985, 51) terms, the truth status of the discourse of science can be seen to be underpinned by the institutional sites from which scientists enunciate their discourse "and from which this discourse derives its legitimate source and point of application (its specific objects and instruments of verification)." The institutional sites that legitimately ground and authorise the discourse of science and the discursive practice of scientific method include the academy and the laboratory.

In marking the importance of institutional sites as critical in legitimating a discourse's truth status, Foucault (1985, 5) elaborates on the importance of asking:

> who is speaking? Who, among the totality of speaking individuals, is accorded the right to use this sort of language (*langage*)? Who is qualified to do so? Who derives from its own special quality, his [or her] prestige, and from whom, in return, does he [or she] receive if not the assurance, at least the presumption that what he [or she] says is true? What is the assurance, at least the presumption that what he [or she] says is true? What is the status of the individuals who—alone—have the right, sanctioned by law or tradition, juridically defined or spontaneously accepted, to proffer such a discourse?

By raising this battery of questions, Foucault brings into sharp focus the matrix of institutional and social relations that are at work in regimes of truth. In other words, Foucault situates the production of truth within relations of power and knowledge that are at all times socially situated. A technology of truth, such as a biometric system of identity verification, emerges, then, as always already *mediated* by a cluster of relations of power and knowledge. As I examine in detail throughout the course of this book, it is precisely the socially mediated status of technology that is repeatedly effaced in so many scientific accounts of biometrics. This leads to claims that biometric systems are to be celebrated because they are objective technologies that remove the biases and prejudices of human observers, and thus deliver impartial and unmediated knowledge of their respective objects of inquiry. Woodward et al. (2003, 254), for example, celebrate what they term the "technological impartiality of [biometric] facial recognition" systems, as, they argue, it is wholly reliant on "objectively measurable facial features" and is "therefore free from many human flaws." This claim, as I demonstrate, is effectively discredited when juxtaposed against the fact that biometric enrolment for some people of colour is sometimes marked by failure as the technology's image acquisition parameters have been based on the white body as the template subject (see Chapter 2).

Situated Knowledges

Despite the flourishing of a number of fields of critical inquiry—including feminist, postcolonial, race, ethnicity, whiteness, poststructuralist and queer studies—that have interrogated and deconstructed the claim that science is an objective and impartial practice, this illusory claim persists across a large body of texts concerned, in particular, with the technological aspects of science. One of the key objectives of this book is at once to critique such claims within the domain of biometric technologies and their social uses, and to proceed to situate these technologies within social and historical formations. As such, I will proceed to read biometric technologies as the products of "situated knowledges" (Haraway 1991, 188). In coining this critical term, Donna Haraway effectively worked to disclose the multiple social and political dimensions that inscribe knowledge production as a result of the embodied locus that the knowledge producer occupies. This embodied locus is marked by the categories of gender, race, ethnicity, sexuality, class, (dis)ability and age. Attempts to transcend all of these qualifying factors simply instantiate gestures of disavowal, precisely as they mark the situated position of those privileged enough not to have to account for their own embodied locus: "Only those occupying the position of the dominators," writes Haraway (1991, 193), "are self-identical, unmarked, disembodied, unmediated, transcendent."

Infrastructural Normativities and Biometrics' "Extrinsic"Categories

Situated in the context of my analysis of biometrics, Haraway's naming of the privilege of the "dominators" to self-represent their knowledge production as unmarked, disembodied and unmediated resonates powerfully with one of the key categories of inquiry that drives my critical approach: *whiteness*. Whiteness studies is a relatively recent field of inquiry. Although it can trace its genealogy back to landmark figures in critical race studies, such as W. E. B. Du Bois and Frantz Fanon, whiteness studies only became consolidated into a recognisable field of inquiry in the 1980s and 90s. This belated consolidation into a formal field of inquiry can be seen to be due to the very power that whiteness has exercised in western culture and its colonial dominions. Although questions of race have been raised and examined in western culture from the seventeenth century onwards, race has been equated, until recently, as "naturally" pertaining to blacks and people of colour. The bulk of theorising and writing on race over the last few hundred years has been conducted by those that Haraway calls "the dominators," in other words, by privileged white subjects. As such, white subjects have positioned themselves as somehow transcending racial inscription and, in effect, as producing what Toni Morrison (1992, 16) has termed "the 'normal,' unracialised, illusory white world." In this narrative of racial disavowal, whiteness has been scripted not as a racial category but, rather, as a universal or template human category that is unmarked by any racial inflections. As Richard Dyer (1997, 3) writes, "At the level of racial representation ... whites are not of a certain race, they're just of the human race." Whiteness emerges, then, as a racial regime that is predicated on invisibilising its own racial status in order to preserve the illusory notion that whites simply and tautologically occupy the unmarked category of the "human." What is produced through this gesture of disavowal is a position of privilege that enables whites to locate, name and identify race everywhere but within and across their own embodied locus. This position of privilege and power is succinctly identified by Ruth Frankenberg (1993, 197): "whites are the nondefined definers of other people." In this schema, whiteness functions to operate as the "unspoken norm" (Frankenberg 1993, 197) against which all "deviations" are identified, measured, evaluated and classified.

In the course of this book, I proceed to track the operations of whiteness as one of the unspoken norms within the discursive practices of biometric technologies and their target subjects. In examining the structuring power of whiteness in the context of biometrics, I argue that whiteness is so *infrastructurally* diffuse in its socio-technological operations as to be imperceptible, whiteness so constitutes the infrastructural fabric of everyday technologies and practices that it cannot appear as a racial category as such (see Chapter 2). In my analysis of infrastructural whiteness, I will be drawing attention to the very *structurality* of its infrastructure; in other words, I want to bring into focus what effectively gets invisibilised when technologies are represented as ideologically neutral "conduits" of data, rather than ideologically inflected constructors of knowledge. In examining the apparent paradox of a racial category, whiteness, that self-represents as unraced, I will argue that it is precisely because whiteness is so enucleated into the material weave of everyday life that one cannot talk of whiteness *as such*. No *as such*, so to speak, because the power of whiteness resides in this capacity to occlude and so mystify its status as a racial category that it too often escapes taxonomic determination, while simultaneously remaining the superordinate racial category that effectively determines the distribution of all other classificatory categories along the racial scale.

In the literature of biometrics, such intersectional categories as race and ethnicity, gender, age, class and (dis)ability are framed as "extrinsic" or "ancillary" to the "primary" identifiers of a subject (for example, the finger in finger-scan biometrics). I argue, on the contrary, that such socially inscriptive categories always already "intrinsically" mark the bodies of subjects who present themselves for biometric processing. This is evidenced by the manner in which a biometrically scanned subject often fails the enrolment process precisely because of her or his gender, race, class and/or disability. Operative across the range of biometric modalities are a series of what I will term *infrastructural normativities* that function to set the image-acquisition standards of the technologies. Normative conceptualisations of race, gender, class, age and (dis)ability are already embedded within the very infrastructure of biometric technologies: they constitute the *a priori* conditions of the technology's operations; their *a priori* status guarantees their invisibility. As such, these infrastructural normativities produce biopolitical effects for those subjects who fall outside their normative parameters. Furthermore, it is precisely the *a priori* status that these infrastructural normativities are invested with that enables the production of such categories as race, gender and so on to be scripted as "extrinsic" or "ancillary" to the primary operations of biometric technologies. Embedded within the very infrastructural operations of the technology are a series of normative presuppositions and inscriptive categories (of race, (dis)ability, gender, age, class) that, because of their normative infrastructural status, cannot be named or rendered visible. Questions of race, gender, (dis)ability, class and/or age are structurally made to signify as "extrinsic" to the operations of the technology as they only come into being, so to speak, when the non-normative subject, as subject of exteriority, comes to enrol in the biometric system and fails because of her or his non-normative status. It is, conclusively, the *a priori*, infrastructural status of a series of inscriptive normativities that enables the advocates of biometrics to declare the technology as impartial, objective and non-discriminatory. What remains unspoken in such pronouncements is the fact that the discriminatory elements of the technology remain unrepresentable because they constitute the *a priori*, infrastructural presuppositions of the technology.

Biopolitics and Disciplinary Normativity

In invoking the concept of infrastructural normativities, I am drawing upon Foucault's notion of disciplinary normativity as a technology of biopower. For Foucault, biopower emerges in the late eighteenth and early nineteenth centuries. Principally concerned with the body and its relation to structures of power, "Bio-power brought life and its mechanisms into the realm of explicit calculations and made knowledge/power an agent of transformation of human life" (Foucault 1990, 143). Biopower, thus, effectively colonises the body, overlaying it with calculatory grids and mathematically inscribing it with formulae that will transform it into an object of knowledge and power. Biopower, Foucault (2007, 1) explains, is "the set of mechanisms through which basic biological features of the human species became the object of a political strategy." Biometric systems, once situated within this conceptual framework, emerge as exemplary technologies of biopower. Biometric systems are predicated on the notion that the body can be subject to economies of explicit calculations that can be mobilised toward political ends. The body, when screened by biometric technologies, is divided into corporeal components (the iris, the fingerprint, the face) that are individually processed and converted into algorithmic formulae and then stored as schematised templates within biometric databases. Individual body parts are, in this process, inscribed within anatomies of biopower that enable the operations of regimes of identification, surveillance and disciplinary normativity. The mathematical and digital transmutation of corporeal attributes, such as fingerprints or facial features, into biometric templates functions to construct biometric technologies as scientifically impartial systems purely driven by the "arithmetic of proof" (Foucault 2003a, 7).

Biometric technologies are capable of scanning the entirety of the body (through gait signature biometrics), its surfaces (facial and finger scans), its depths (vein recognition biometrics) and its chemical emanations (odour recognition biometrics), "making it possible," in biopolitical terms, "to bring the effects of power to the most minute and distant elements. It assures an infinitesimal distribution of the power relations" (Foucault 1982, 216). This microphysical distribution of power relations is played out through the manner in which biometric technologies capture minute body parts (for example, the iris), convert them to digital templates and then proceed to store them within interoperative, multimodal and transnational electronic databases that connect, for example, the iris scan of an Iraqi citizen captured by the U.S. military to globalised systems of identification and surveillance (see Chapter 3). In biopolitical terms, that is, in terms of the governance, surveillance and regulation of target populations, biometric technologies function as systems of "centralized individualization" (Foucault 2006, 49): they individuate the biometrically scanned body while simultaneously inserting the schematised template into databases that enable the centralised monitoring of the individual, including the monitoring of their movement across local, national and transnational locations and spaces.

In his critical mapping of the emergence and consolidation of the surveillance society, David Lyon (2005, 1) deploys the term "surveillance as social sorting" in order to highlight the political dimensions of this configuration of bodies, technologies and practices of surveillance: "surveillance today sorts people into categories, assigning worth or risk, in ways that have real effects on their life-chances. Deep discrimination occurs, thus making surveillance not merely a matter of personal privacy but of social justice." As Lyon makes clear, technologies of surveillance cannot be seen as merely neutral technologies that transcend political relations of bodies and power; on the contrary, such technologies must be seen as performing more than automated procedures of identification: they are instrumental in evaluating embodied subjects and in allocating them within hierarchised positions of power. Jane Caplan and John Torpey (2001, 3), in their tracing of the development of practices documenting individual identity in the modern world, disclose the effaced biopolitical dimensions that fundamentally inscribe identificatory technologies and their practices: "The question 'who is this person?' leaches constantly into the question 'what kind of person is this?'"

As I proceed to demonstrate throughout this book, the identificatory procedures of biometric technologies are fundamentally inscribed by tacit moralising assumptions, normative criteria and typological presuppositions. So, for example, in analysing the operations of gait signature biometrics, designed to identify a prospective criminal or terrorist by the way they walk, I proceed to discuss the moralising assumptions and normative criteria that must be used in order to enable the differentiation between the normative walk of the citizen-subject and deviant walk of the suspect subject. Encoded within this biometric system of identification, I argue, is a cluster of racialised and able-bodied assumptions about what constitutes a normative, non-criminal walk. The foundational role that normative conceptualisations of the body play in biometric systems is perhaps most graphically demonstrated by the complete elision of people with disabilities from biometric failure-to-enrol rates, as the image acquisition parameters of the technology have been predicated solely on able-bodied subjects. Within the normative schemata of biometric technologies, people with disabilities are doubly invisibilised: first, through the phenomenon of failure to enrol and, second, through their systemic elimination from the failure-to-enrol rate of a particular biometric system (see Chapter 3). What is operative here is precisely what Lyon terms "deep discrimination," in which normative conceptualisations of the body determine the operational infrastructure of a technology that, in turn, effectively discriminates against those who fail to conform to its operating (normative) "standards."

These normative assumptions function to situate gait signature biometrics, for example, within biopolitical schemata that intertwine biotypological conceptions of the body (in which bodies are assigned the identificatory label of "types") with a disciplinary anthropometrics (in which the measurement of an aspect of the body in question is juxtaposed against an understanding of the "normal" and "deviant" body). This intertwining of assumptions about body types with disciplinary regimes of normativity is not something that is unique to biometric technologies. As Jane Caplan (2001, 51) argues in her historical analysis of the emergence of protocols of identification in nineteenth-century Europe,

> in virtually any systematics of identification, everyone is not only "himself" [or "herself"] but also potentially the embodiment of a type, and in an important respect the history of identification is a history not so much of individuality as of categories and their indicators.

Biometrics' Genealogies

Caplan's marking of the critical manner in which identificatory technologies and practices are always already marked by a series of typological categories and their embodied indicators is graphically evidenced by an historical survey of the field of biometrics. In Chapter 1, I stage a tracking of the historical conditions of the emergence of biometric technologies. In order to make visible the complex genealogy of technologies that has informed the development of contemporary biometrics, I deploy an enlarged, historical sense of the term *biometrics*. In this book, biometrics does not only refer to contemporary, computer-automated technologies that authenticate or identify enrolled subjects. Rather, taking the term in its most literal sense, it refers to a cluster of technologies that have all been preoccupied with the measurement of the body in order identify, classify, evaluate and regulate target subjects. My genealogical account of biometrics, then, includes an entire series of precursor technologies that encompasses physiognomics, phrenology and anthropometry. In the course of my genealogical investigation, rather than focusing on a type of originary moment of emergence of biometrics (for example, the development of the biometric technology of fingerprinting in the nineteenth century), I discuss the diffuse and complex history of technologies of body measurement and individual identification and its attendant enmeshment within biopolitical assumptions about race, gender, class and (dis)ability.

In staging this historical analysis, I do not propose an uninterrupted line of continuity between these past biometric technologies and today's biometric systems. Rather, in deploying what Foucault (1986, 146) terms, after Friedrich Nietzsche, a "genealogical method," I attempt to demonstrate both the points of connection between past and contemporary biometric technologies *and* the points of rupture and discontinuity. Foucault (1986, 146) describes the genealogical method thus:

> Genealogy does not pretend to go back in time to restore an unbroken continuity that operates beyond the dispersion of forgotten things; its duty is not to demonstrate that the past actively exists in the present, that it continues secretly to animate the present, having imposed a predetermined form to all its vicissitudes. Genealogy does not resemble the evolution of a species and does not map the destiny of a people. On the contrary, to follow the complex course of descent is to maintain passing events in their proper dispersion; it is to identify the accidents, the minute deviations—or conversely, the complete

> reversals—the errors, the false appraisals, and the faulty calculations that gave birth to those things that continue to exist and have value for us.

In proceeding to situate, in Chapter 1, biometric technologies within a genealogy of anthropometric and identificatory technologies and practices, I attempt to trace a line of "descent" that bifurcates, fractures, reconnects and deviates from the present. So, for example, in establishing the system of relations that holds between nineteenth-century phrenology and a contemporary technology such as No Lie MRI (see Chapter 5), I draw attention to the elements of continuity and discontinuity that inscribe both systems. Nineteenth-century phrenology was premised on the belief that a person's moral character was indelibly inscribed in the topographical specificity of the brain, and that cerebral locationism would enable a diagnosis of the qualities of the brain, including the cerebral location of various moral qualities such as lying, conscientiousness and secretiveness. This concept of cerebral locationism can be seen actively to inform a contemporary technology such as No Lie MRI. No Lie MRI claims to have isolated specific locations in the brain that are activated whenever a subject tells a lie. This relation of continuity between phrenology and No Lie MRI is, however, also fractured by a fundamental difference. Whereas the success of nineteenth-century phrenology was premised on the divinatory powers of the phrenologist-specialist to provide a diagnostics of the brain in question, with No Lie MRI, the success of the diagnostic process is entirely premised on the "objective" operations of the scanning technology; the technology is touted as single-handedly providing the scientific evidence, thus relegating the scientist to a secondary position as the mere "conduit" who merely relays the extracted information.

In arguing that "to follow the complex course of descent is to maintain passing events in their proper dispersion," Foucault (1986, 147) problematises western historiography's reliance on a teleological understanding of history. Within teleological understandings of history, the past is seen as supplying the archaic and primitive ground for categories, technologies and practices that, with the progressive unfolding of time, develop and achieve their proper and sophisticated forms. In this teleological schema, nineteenth-century phrenology, with its belief that one could diagnose a person's character according to the specific topography of a subject's brain, is dismissed by today's science as an example of "pseudo" or "junk" science, in contradistinction to the "proper" or "real" science of contemporary practice. In deploying the genealogical method, I refuse to make this historically untenable distinction. By situating scientific technologies and practices within the locus of their historical dispersions, I argue what is now considered "junk" science was then considered "proper" science and that consequently, it was invested with a discursive authority that produced material social effects.

The material social effects of a scientific practice such as phrenology (a practice that is now thoroughly discredited in the domain of science) become clear when examined in the context of the famous nineteenth-century criminal anthropologist Cesare Lombroso (see Chapter 1). Lombroso's criminal anthropology effectively informed the policies and practices of Italian policing and criminal justice (Gibson 2002, 127–74). His criminal anthropology gave shape to regimes of criminal profiling that were raced, gendered and classed; and it determined who became a target subject within regimes of preventive criminology. It is in this context that the genealogical method can offer invaluable insights into the occluded historical dimensions and relations that hold in the present.

In his discussion of the contemporary surveillance society, Lyon (2003, 15–16) argues that one of its distinguishing characteristics is "the trend towards attempted prediction and pre-emption of behaviors, and of a shift to what is called 'actuarial justice' in which communication of knowledge about probabilities plays a greatly increased role in assessment of risk." Even as contemporary regimes of criminological prediction and pre-emption are, as I discuss in Chapters 4 and 5, reinvented in such technologies as Project Hostile Intent and No Lie MRI, they actually find their conditions of emergence in the nineteenth century through the development of the knowledge/power/body matrix that, as discussed earlier, Foucault named biopower.

Hubert Dreyfus and Paul Rabinow (1982, 134) offer a lucid exposition of biopower's historical conditions of emergence:

> In Foucault's story, bio-power coalesced around two poles at the beginning of the Classical Age. These poles remained separate until the beginning of the nineteenth century, when they combined to form the technologies of power which still recognizably characterize our current situation.

The "two poles" that come together at the beginning of the nineteenth century are a "concern with the human species" and an acute focusing of attention on "the body" (Dreyfus and Rabinow 1982, 134). The concern with the human species sees "scientific categories ... rather than juridical ones bec[o]me the object of political attention in a consistent and sustained fashion" (Dreyfus and Rabinow 1982, 134). This concern with the human species generated the development of an array of disciplines devoted to examining, quantifying, diagnosing and predicting demographics, population movements, disease and epidemics, criminal "classes," "types" and so on. Concomitantly, the development of these new scientific disciplines established the body as the focal point of these practices of inquiry. The body became the object to be measured, gauged, investigated, probed, mapped and manipulated. As object of inquiry, the body was submitted to the clinical and technocratic gaze of *regimes of visuality*, which proceeded to assess, diagnose, identify and put to use its various properties.

Visual Regimes and Optics of Power

In drawing upon the theoretical term *visual regimes*, my aim is to problematise the very practice of seeing. Seeing, as an embodied practice, would appear to be a wholly physiological and cerebral phenomenon. One looks with one's eyes and what one sees is immediately and naturally processed by an instantaneous combination of physiological and cerebral processes. Taking my cue from Foucault's theorisation of the fundamental ways in which such seemingly natural embodied practices, such as sexuality, are always already discursively mediated by relations of knowledge/power, I transpose this nexus of knowledge/power to the practice of seeing. Seeing, in the process, emerges not as a purely physiological and cerebral process but, rather, as a socially mediated and historically situated phenomenon inflected by relations of power. In coining the term "regimes of vision," Jonathan Crary (1998, 3), in his work on the historicity of vision, brings into focus the "problematic phenomenon of the observer":

> For the problem of the observer is the field on which vision in history can be said to materialize, to become itself visible. Vision and its effects are always inseparable from the possibilities of an observing subject who is both the historical product *and* the site of certain practices, techniques, institutions, and procedures of subjectification.

In locating his analysis within the phenomenon of the observer, Crary proceeds to mark both the embodied nature of seeing and the situated and mediated status of the subject doing the seeing or observing. The practice of seeing therefore emerges as a physiological process that is historically shaped and socially mediated by particular technologies, knowledges and discursive practices of subject formation. The discursive shaping and mediation of the practice of seeing is perhaps most effectively captured by the concept of the "gaze." Developed within the field of visual culture, the concept of the gaze refers to the manner in which the practice of seeing is fundamentally inflected by such embodied categories as gender, race, sexuality, class and (dis)ability. A number of visual culture theorists have demonstrated how what the observer sees is inscribed by relations of power that, through the gaze, often establish subject and object positions. For example, in her landmark work on the male gaze, Laura Mulvey (1999 [1975], 383) posited that:

> In a world ordered by sexual imbalance, pleasure in looking has been split between active/male and passive/female. The determining male gaze projects its fantasy onto the female figure, which is styled accordingly.

> In their traditional exhibitionist role women are simultaneously looked at and displayed, with their appearance coded for strong visual and erotic impact so that they can be said to connote to-be-looked-at-ness.

Mulvey proceeds to map the dimensions of the male gaze across the wide spectrum of cultural production—from film to advertisements—in order to evidence her thesis. Subsequent work on the male gaze, however, proceeded to question some of the problematic assumptions that underpin Mulvey's thesis: for example, that there is a "unified masculine model of spectatorship" (Stacey 1999, 391) and that female spectators exercise modalities of agency that Mulvey fails to acknowledge. The homogenising notion of a unified male gaze has been problematised and complicated by the work, for example, of bell hooks (1992, 115–31), who has mapped the racialised dimensions of the gaze in the context of African American culture and white supremacism. Similarly, the male and female gazes have been shown to be differentially constituted by spectatorships and observer positions marked by queer desire that effectively recode and reconfigure the heterosexist dyad: male/female, active/passive (Mercer 1994, 171–219).

In my analysis of biometrics in the context of regimes of visuality, I address the ongoing reproduction, within the scientific literature, of what Foucault terms the "clinical gaze." In *The Birth of the Clinic*, Foucault (1975, xiii) maps the discursive forces instrumental in the formation of medical perception within the space of the clinic. The clinical gaze, Foucault argues, is underpinned by a "medical rationality [that] plunges into the marvelous density of perception, offering the grain of things as the first face of truth, with their colors, their spots, their hardness, their adherence." The clinical gaze, he writes, is constituted by "empirical vigilance receptive only to the evidence of visible contents. The eye becomes the depositary and source of clarity; it has the power to bring a truth to light that it receives only to the extent that it has brought it to light" (1975, xiii). Foucault here delineates the coextensiveness of the clinical gaze with the position of a disembodied, unsituated observer who is represented as merely delivering data or knowledge gleaned from the operations of a particular technology in a purely unmediated manner. Operative in the clinical gaze are "the privileges of a pure gaze, prior to all intervention and faithful to the immediate, which it took up without modifying it" (Foucault 1975, 107).

As I demonstrate throughout the course of this book, the clinical gaze, as a pure gaze that merely delivers up the "truth" of its subject of inquiry, pervades the field of biometric literature. In Chapter 5, for example, I discuss the new technology of Brain Fingerprinting. This technology is being promoted as being able to identify a criminal by subjecting her or his brain to an electroencephalographic (EEG) scan. This scanning technology promises to take the observer beyond the surface duplicities and feints of the suspect subject. Penetrating into the inner sanctum of the subject, the brain, the clinical gaze, augmented by visualising technologies, pierces the opaque morphology of skin and bone in order to deliver up the "truth": whether or not the subject has committed a crime will be revealed by the fact that this technology is purported to detect either the presence (guilty) or absence (innocence) of memories of the criminal act. Furthermore, the technology is presented as delivering this information independently of any human bias, observational or otherwise:

> The entire Brain Fingerprinting system is under computer control, including the presentation of the stimuli, recording of electrical brain activity, a mathematical data analysis algorithm that compares the responses to the three types of stimuli ["targets," "irrelevants," and "probes"] and produces a determination of "information present" or "information absent," and a statistical confidence level for this determination. At no time during the analysis do biases and interpretations by the person giving the test affect the presentation or the results of the stimulus presentation.
>
> (Brain Fingerprinting Laboratories)

Operative here is a fusion of the clinical gaze, the process of diagnostic analysis and the constitutive practice of interpretation. The series of mediating processes (including linguistic, scopic and technological) that enable the process of diagnostic analysis is occluded by the claim that what is revealed is, in fact, purely computer driven and independent of the observing subject who will necessarily interpret the data. This discursive effacement of the observer can be seen to be a structural effect of digital technology that, as Crary (1998, 1) argues, effectively liquidates the constitutive role of the observer in the production and interpretation of digital images precisely because these images do not reproduce the mimetic representations of analogue technologies such as film and photography. The recession of mimesis from the visual plane of the observer, and its replacement by computer fabricated digital images, functions to relocate "vision to a plane severed from a human observer":

> Most of the historically important functions of the human eye are being supplanted by practices in which visual images no longer have any reference to the position of an observer in a "real," optically perceived world. If these images can be said to refer to anything, it is to millions of bits of electronic mathematical data. Increasingly, visuality will be situated on a cybernetic and electromagnetic terrain where abstract visual and linguistic elements coincide and are consumed, circulated, and exchanged globally.
>
> (Crary 1998, 1–2)

The liquidation of the mimetic effect in the visual field of digital images functions to produce the discursive effect of imaging technologies that are seemingly free of human bias. The digital image of brain waves produced by Brain Fingerprinting, for example, does not correlate to the optically perceived "real" world of the observer. However, the observer "returns," so to speak, in the after-moment of digital image production, precisely in order to make sense of and interpret the digital image. This "return" of the hermeneutic observer, as constitutive interpreter/mediator of the digital image, is precisely what is occluded by the non-mimetic status of the digital image when it is represented as structurally independent of a human observer. This "return" of the hermeneutic observer, moreover, must be seen as a structuring *a priori* that at once precedes the "reading" of the image in question, even as it constitutes its conditions of cultural intelligibility. As I discuss in some detail in Chapter 5, what the seemingly objective scientific gaze claims to "discover" or "disclose" in the purity of its objective field of perception is, structurally, predetermined to appear as such through the (effaced) analytical and discursive conditions that enable the very *re-cognition* of what is seen.

In my deployment of the term *visual regimes* in this study of biometrics, my focus on the mediated status of the practice of seeing will be informed by the concept of regulative power encoded in the word "regime." A visual regime, as discursive construct, regulates reproducible and delimited ways of seeing that are necessarily enmeshed within technologies of power. "Vision requires," Haraway (1991, 193) writes, "instruments of vision; an optics is a politics of positioning. Instruments of vision mediate standpoints." In other words, visual perception is critically dependent on *technologies of seeing*, both "hard" (for example, cameras and scanners) and "soft" (such discursive practices as the clinical or technocratic gaze).

The optics generated by such technologies of vision are, in turn, indissociably inscribed with relations of power that mediate subject and object relations. In Chapter 3, for example, in my analysis of the U.S. military's enforced biometric scanning of Iraqi citizens' irises in order to identify terrorists, I attempt to materialise the manner in which technologies of vision are inscribed with relations of biopower. The enforced biometric scanning of Iraqi citizens by the U.S. military is predicated on violently unequal relations of power, specifically, on subject/object relations. Situated within the context of the imperial conduct of war in Iraq, this militarised biometric practice represents the exercise of invasive power and control by the U.S. military over the bodies of Iraqi citizens. Iraqi citizens no longer have autonomy or control over their own bodies, more specifically, over the biometric identifiers of their own bodies. Positioned within this militarised visual regime, they are figured as mere objects that must make available their bodies to the superior force. Represented as coextensive with the imperially occupied terrain of Iraq, they are compelled to submit to a systemic exercise of biopower that mines their biometric signatures. In this militarised context of regimes of visuality, the optics of power emerge as instrumental in the maintenance of imperial rule and violent subjugation;

and, in this charged context, Haraway's (1991, 192) acute observation resounds: "Vision is always a question of the power to see—and perhaps of the violence implicit in our visualizing practices. With whose blood were my eyes crafted?"

Biometrics in the Service of Colonialism and Empire

Haraway's provocative question—"With whose blood were my eyes crafted?"—can be historically answered by tracing the very development of biometric technologies within the violent fields of western colonialism and imperialism (see Chapter 1). Traditional fingerprint biometrics, based on the acquisition of a subject's inked fingerprint impression and the consequent classification of this impression within a taxonomic filing system, was first developed by the colonial British administration in India. As Simon Cole (2002, 63) demonstrates, this biometric technology was developed "in response to the problem of administering a vast empire with a small corps of civil servants outnumbered by hostile natives." Cole notes how the colonial system of fingerprint identification emerged in response to an uprising in 1857 by Indian conscripts against the British, resulting in the temporary control of Delhi. "The Mutiny," writes Cole (2002, 64), "heightened 'the need to enforce law and order in the unruly colonies more severely,'" and, at the same time, "the need to reinforce a sense of Britain's proper role in history as a beacon of order and civilization in a world of darkness and barbarism." The technology of fingerprint biometrics was developed in order to construct a colonial system of identification and surveillance of subject populations in the face of British administrators who "could not tell one Indian from another" (Cole 2002, 64). Inscribed in the incipient moment of fingerprint biometrics' development is a racialised agenda driving the system's mode of identification, its visual regime of surveillance and its biopolitical practices of colonial administration.

The colonial administrative uses of biometrics in the service of empire and the role of the sciences in developing apparatuses for the classification, monitoring, surveillance and disciplining of target populations are not things that can be merely relegated to the past. On the contrary, as I demonstrate in Chapter 3, the deployment of biometric technologies in the service of empire is very much the business of contemporary biopolitical practice. I evidence this by discussing the development, by the U.S. Department of Defense Biometrics [DoDB], of "identity dominance" through biometric technologies. In the context of the war on terrorism, identity dominance signals the deployment of biometric technologies in order to "link an enemy combatant or similar national-security threat to his [or her] previously used identities and past activities, particularly as they relate to terrorism and other crimes" (Woodward 2005, 30). The emergence of this new term, "identity dominance," establishes the assimilation of biometric technologies into the biopolitical configuration of U.S. empire, war and bodies. The contours of this matrix of biometrics, empire, war and bodies are marked by an approach that is described by the DoDB as "multitheater, multiservice, multifunctional, and multibiometric." I analyse the U.S.'s deployment of gait signature biometrics, in order to identify prospective terrorists, and iris scan, in the context of the war in Iraq, as a way of achieving identity dominance, and discuss the biopolitical ramifications of this.

In his lectures on biopolitics, Foucault (2008, 67) addresses the manner in which, in western societies, "liberalism ... and disciplinary techniques are completely bound up with each other." "[L]iberalism," Foucault (2008, 66, 72) argues, is "predicated on the interplay of security/freedom," to the point that freedom must be seen as "the correlative of apparatuses of security." It is in this context that Anthony Burke (2001, xxxiv) argues for security to be conceptualised as a "political technology" constitutive of "a network of practices and techniques which produce and manipulate bodies, identities, societies, spaces and flows." As I discuss in Chapter 3, a complex array of biometric apparatuses of security are being developed by western governments in order to secure liberal notions of freedom through "identity dominance." This mobilisation of biometric technologies in order to secure identity dominance can be seen to be moving into the realms of science fiction through the funding and development, by the U.S. Department of Homeland Security, of Project Hostile Intent (PHI). PHI will be based on the deployment of a variety of scanning technologies that are designed to sweep across a crowd of people in order to detect signs, such as increased heart rates or perspiration, that indicate the presence of a prospective criminal subject with "hostile intent." Coextensive with the sci-fi realm represented by the Hollywood film *Minority Report*, the goal of PHI is to identify a prospective criminal in advance of the fact of their having committed a crime. With PHI, I argue, the target body is enmeshed within a regime of biopower designed to bring its biological processes and attributes into the realm of explicit calculations. Pulse rate, perspiration and micro facial expressions are all digitally calibrated and gauged against integrated scores that are supposed to indicate the hostile intent of a subject predisposed to criminality. Presupposed in this integrated score of "malfeasance likelihood" is a disciplinary norm that establishes the guiding templates for "correct," "normal" and "appropriate" behaviour against which deviations signal targets that pose security risks.

Somatechnics: Bodies, Technologies

In tracing the emergence of biopolitics, and its new forms of "political rationality," Foucault marks a significant shift from juridical to administrative power:

> The first principle of this new rationality was that the state, not the laws of men or nature, was its own end. The existence of the state and

> its power was the proper subject matter of the new technical and administrative knowledge, in contrast to juridical discourse, which had referred power to the other ends: justice, the good, or natural law ... The object to be understood by administrative knowledge was not the rights of the people, not the nature of divine or human law, but the state itself ... And this required the gathering of information on the state's environment, its population, its resources and its problems.
>
> (Dreyfus and Rabinow 1982, 137).

The gathering of information on the state's population engendered the development of a series of biometric technologies that proceeded to subject bodies to various forms of mathematical measurement, classification and surveillance. To this end, I discuss in Chapter 1 a panoply of anthropometric systems that were developed in the nineteenth century, including inked fingerprinting, craniometry, osteometry, somatometry and so on. These anthropometric systems were increasingly, in the face of the development of a biopolitics of administrative power, coopted and deployed by the police, colonial administrators and their various functionaries. At the centre of all the anthropometric and biometric systems that I discuss, in the course of this book, is the human body. Biometric technologies are systems fundamentally concerned with the technological measurement of the body. As such, conceptual categories of the body and technology constitute biometrics' very conditions of possibility, that is, if there is no body, then there can be no biometric as such. I mark this seemingly obvious point precisely in order to begin to problematise and flesh out, as it were, "the body" as a category that, throughout much of the biometric literature that I analyse, retains a wholly self-evident status.

In contradistinction to a conceptualisation of the body as a purely natural datum, I proceed to conceptualise the body as an entity that can only achieve its cultural intelligibility as "body" precisely because it is always already inscribed by a series of discursive and technological mediations. As I have discussed in detail elsewhere, I understand the body as indissociably tied to technology and not as something that stands in contradistinction to it (Pugliese 2005a). Drawing on the work of Jacques Derrida, I refuse the binary: body/technology or natural/synthetic. In his theorisation of the relation between body and technology, Derrida (2002a, 244) articulates the inextricable tie between the natural ("*physis*") and the synthetic ("*technè*"). Derrida (2002a, 244) emphasises that this relation "is not an opposition; from the very first there is instrumentalization [*dés l'origine il y a de l'instrumentalisation*] ... a prosthetic strategy of repetition inhabits the very moment of life. Not only, then, is technics not in opposition to life, it also haunts it from the very beginning." From the very beginning, then, the body is always already intextuated and instrumentalised by a series of technologies. At the most elementary level, this process of technical inscription, through the technology of language, is essential in rendering the body culturally intelligible. This understanding of the body as something always already inscribed by technology is succinctly encapsulated by the neologism *somatechnics*. Coined by Susan Stryker and developed conceptually by a number of us at the Department of Critical and Cultural Studies, Macquarie University, somatechnics concisely configures this understanding of the inextricable relation between bodies and technologies (see Somatechnics Research Centre; Pugliese and Stryker 2009).

Across the biometric literature that I discuss, the body is viewed as central to the entire conceptual apparatus of biometric systems, precisely because biometrics is predicated on the technological reading and measurement of the body. Yet, throughout this literature, the body is always represented as a purely biological datum that signifies self-evidently. Throughout this book, I call the body into question in order to begin to mark its historically located and discursively inscribed status. As such, I track the body in the context of biometric technologies as a varied and historically shifting entity rendered culturally intelligible by the specificities of the technologies that mark and configure it and that, in turn, it shapes and inflects. Moreover, in calling into question the given status of the body as a purely self-identical biological datum, I proceed to deconstruct the philosophical presuppositions that are constitutive of biometric understandings of the body, even as they remain effaced and unquestioned.

In Chapter 4, for example, I bring into critical focus the metaphysical presuppositions that underpin biometric understandings of the body as absolutely unique in the identity of its self-presence. Across much of the relevant literature, biometric technologies are presented as providing virtually foolproof systems of identification and/or verification of the subjects it screens. Yet, despite these claims, the literature also acknowledges that biometric systems are open to being tricked by fraudsters. In order to counter the tactics used by fraudsters to "fool" biometric systems, biometric scientists and technologists are in-building within the technologies a number of tests designed to detect fraudsters. One of the key fraud detection methods being deployed by biometric systems is so-called "liveness testing." Liveness testing is being used to determine whether the person being screened by the system is actually present (and "alive") rather than a simulacrum (for example, a video recording) reproducing a stolen identity.

In the course of Chapter 4, I proceed to situate biometric liveness testing within a Derridean critique of the metaphysics of presence in order to disclose the unacknowledged philosophical dimensions that inform scientifico-empiricist understandings of the body, technology and identity. As a subject's signature, biometric or otherwise, must at once be unique *and* reproducible in order for it effectively to operate within identity recognition systems such as biometric technologies, it immediately opens up the possibility for someone else to mimic and reproduce the unique identity signature of a particular subject. In their absolute dependence on the figure of a subject's biometric *signature*, biometric technologies are, I conclude, permanently open to the possibility of citational grafts and identity frauds. In the context of the increasing ensnaring of the body within biopolitical regimes of surveillance, security and control, my deconstruction of the metaphysics of presence that underpin empiricist notions of the body opens up the possibility to account for practices of agency that subvert and mock totalising claims about biometric systems' foolproof fraud detection capabilities. Situated within a Foucauldian (1980, 142) understanding of power, in which power is not simply a purely negative force but simultaneously productive of in-built resistances—"there are no relations of power without resistances"—these practices evidence the instability and gaps within biometric systems. They evidence, furthermore, a microphysics of power that haunts all techno-scientific attempts to construct an infallible biometric system that cannot be breached by fraudsters and hackers.

In placing under critical question the concept of the body as a purely biological datum, I simultaneously interrogate the concept of technology as a given that is always ready to hand; that is, I question positivist understandings of technology by focusing on the complex ways in which technology is constituted by the discursive practices that produce it and enable the uses to which it can be put. Throughout this book, I conceptualise technology in the broadest sense of the word. Technology is not only hardware (scanning technologies or computers), but also the complex array of cultural technics that are mobilised in the process of sociocultural life; this includes such "intangible" things as language and mathematics. Language is a fundamental human technology that is deployed to name, identify, inscribe and categorise the world. It acts upon and discursively shapes the world and its objects, rendering them intelligible as identifiable things. I draw attention to language as a mediating technology, as language's constituting materiality is precisely what gets repeatedly effaced in empiricist discourses of science and technology. In my analysis of No Lie MRI, for example, which purports to be able to scan the brain of a subject in order to produce images that will disclose whether or not she or he is lying, the image of a "lie" captured by No Lie MRI is presented in the literature as having been produced without any mediation (see Chapter 5). No Lie MRI claims that the technology "provides unbiased methods for the detection of deception" and that it "represents the first and only direct measure of truth verification and lie detection in human history!" (No Lie MRI). No Lie MRI's image of the "lie" is accomplished through a series of technological and cultural mediations: algorithmic, digital, visual and hermeneutic. Yet it is precisely these interconnected levels of mediation that must be invisibilised and effaced in order to produce the illusion that the technology delivers its truth of the lie in an unmediated manner.

Biometric Proxies and the Reconfiguring of Subject, Body and Identity Relations

The authenticating and identificatory logic of biometric systems is, as I discuss in Chapter 3, predicated on generating a template proxy of the subject; so, for example, an iris scan, transmuted into a biometric template, operates as a proxy for the biometrically enrolled subject every time he or she attempts to access a particular site or database. Encoded in this process, however, is a series of contradictory, if not aporetic, effects that problematise liberal-humanist conceptualisations of both identity and the subject. The aporetic logic of citationality, whereby the truth-status of a subject's re-enrolling template is adjudicated precisely by its failure exactly to coincide with the original enrolment template, inscribes univocal conceptualisations of identity and the subject with a heteronomous law of the self-same-as-other. Indeed, the very status of the key signifiers of biometric identification and verification—uniqueness, authenticity and veridicity—are predicated on an unacknowledged dependence on the other: the self-same subject must generate a micrological series of citations-as-differentiations that de-totalise her or his identity, even as these citations-as-differentiations function to affirm the seeming univocality of identity. Operative here, in effect, is an aporetic logic within which a "true" and "authentic" identity must be simultaneously, at the time of re-enrolment, non-identical to itself—in other words, as I explain in some detail in Chapter 4, at some minimal level, the re-enrolling scan must appear as a "fraud" in relation to the original enrolment template.

The western legal category of the subject is founded on the Enlightenment conceptualisation of identity as fixed, unchanging and self-same. Biometric systems of identification and verification are predicated on this Enlightenment understanding of the subject: the authorising logic of these systems is driven by the notion that, despite micro permutations, the empiricity of flesh (the iris, the face or the fingerprint) encodes an identity that is continuous or identical with itself throughout the subject's existence. Yet, as I argue in the course of Chapter 4, within biometric systems this logic must be underpinned, simultaneously, by a dissemination and decentring of self-same identity. This other, heteronomous logic is perfectly encapsulated in the following biometric formula: "Identification is often referred to as *1:N (one-to-N* or *one-to-many*), because a person's biometric information is compared against multiple (N) records" (Nanavati et al. 2002, 12). One-to-N or one-to-many names the manner in which the unitary identity of the subject is already invested with the law of the other (signatory citation as different in every instance). It is this heteronomy that guarantees the conditions of possibility for the self-same to be constituted as an identifiably unique identity, even as it opens up the same to a movement of discontinuity and dissemination (across various institutional sites and biometric systems with every instance of re-enrolment).

In the course of this book, I proceed to analyse this movement of discontinuity and dissemination of identity in the context of biometric technologies by theorising biometric enrolment templates in terms of scattered "body-bits." Biometric enrolment templates are not, effectively, body-bits at all: they contain no biological material; rather, they algorithmically transmute corporeal features into digital data. However, this digital data is nothing if not a tropological version of a subject's body-bits: no fleshly body, no biometric template; decorporealised body-bits are predicated on the corporeality of the subject's "distinctive physiological characteristics" (Nanavati et al. 2002, 10). The very movement of translation from one to the other produces the rhetorical turn of tropology: biometric templates are tropic proxies of the body, specifically, they are synecdoches of the subject (see Chapter 2). When theorised in terms of their constitutive rhetoricity, the biometric production of templates reproduces the figural logic of synecdoches: that is, every biometric template functions synecdochically in terms of a part signifying the larger whole of the enrolled subject. Biometric proxies, I argue, are invested with juridico-political dimensions as they operate as synecdoches of the legal category of the subject. In other words, when confronted by a biometric system, unless one is able to produce a template, one is directly denied the subject status of legal personhood; whether or not a subject is enabled to take up this position directly determines whether or not they may be given legal or authorised access to restricted space and/or information.

Biometric systems operate on a splitting of identificatory body-bits (and I use the term "bits" in order to underline the digitised nature of corporeal data) through the production of proxies of the subject. This, in turn, opens up a fundamental scission between the subject and the exercise of agency over their very body and their identity. These biometric proxies as digitised body-bits, moreover, are effectively inscribed within national and globalised interoperative databases. Within this economy of dispersed body-bits, the relationship between the subject and their exercise of agency over their body is fissured. For example, against either the intention or will of the subject, a subject's body-bit, once converted into a biometric template, will effectively disclose the identity of the clandestine or dissimulated subject. Hostage to the timbre of its voice and the colour of its irises, the body offers itself up despite the subject. As Irma van der Ploeg (2005, 113) observes, these biometric signs "turn the individual's machine-readable body into a witness against themselves."

I pursue, in Chapter 3, the critical transformations at work in the biometric dissemination of identificatory templates and the consequent scission between body, subject and identity. Biometric templates exemplify the post-biological, *in silico* networking of a subject who can no longer govern or control the heterogeneous dispersal of their identificatory body-bits as biometric template-proxies. It is at this juncture that a disarticulation is enunciated, a disarticulation that vitiates a subject's control over their heterogeneous body-bits and identity proxies. At this juncture, a subject's biometric proxies may be mobilised, indeed, as agents of the state deployed to ensnare and convict the targeted individual. Within these political economies of biopower, biometrics interweaves flesh with algorithms in order to archive body-bits that, with the click of a mouse, reconstitute identity regardless of a subject's permutations. At work here are the biopolitical operations whereby the subject is riveted to the fatality of the biological, even as the biological is transmuted *in silico*. Biometrics must be seen, I conclude, as a technology that is instrumental in the exercise of contemporary biopower and its demand for the production of "docile bodies." In his mapping of the historical emergence of biopower in nineteenth-century Europe, Foucault (1982, 138) notes that:

> The human body was entering a machinery of power that explores it, breaks it down and rearranges it. A "political anatomy," which was also a "mechanics of power," was being born; it defined how one may have a hold over others' bodies, not only so that they may do what one wishes, but so they may operate as one wishes, with the techniques, the speed and efficiency that one determines. Thus discipline produces subjected and practised bodies, "docile" bodies.

Biometric technologies exemplify these biopolitical operations on the body: they machine the body, breaking it down to identificatory, *in silico* body-bits and rearrange it into template proxies that are networked across multiple institutional sites and databases. Biometric systems are instrumental in the reproduction and maintenance of a political anatomy of the body premised on a mechanics of power that splits and fragments the hold between body, subject and identity. In this schema, the supposed fundamentals of liberal democratic societies—the individual and her or his autonomy, privacy, free will and control over their body—emerge as illusory categories underpinned by disciplinary regimes of subjection instrumental in the production of docile bodies. In the context of my analysis of such technologies as PHI, Brain Fingerprinting and No Lie MRI, I delineate this biopolitical production of docile bodies that will offer up the "truth of the body" regardless of the subject's will (see Chapters 3 and 5). I conclude by tracing the ramifications of biometric futures in the context of contemporary biopolitical relations of power and their target subjects.

1 A Genealogy Of Biometric Technologies

Biometric Prehistories

As with all technologies, biometrics did not suddenly emerge from a singular and easily identifiable point of origin. Rather, contemporary biometric technologies are inscribed within a complex history that intertwines social, political and technological factors. In her work on facial recognition biometrics, Kelly Gates (2005, 39), for example, tracks a number of "key nineteenth-century developments in identification systems [that] evidence a perceived need to use the body itself as a marker of identity at the time of modern state expansion." The genealogy of biometrics that I proceed to outline is constituted by systems of relations, differences and end-points. As discussed in the Introduction, in this chapter I will be drawing on Foucault's genealogical method in order to proceed to map the varying precursor technologies of body measurement and individual identification, including physiognomics, phrenology and other forms of anthropometrics that preceded the emergence of contemporary biometrics. While such technologies might appear quaint and entirely redundant in the contemporary context, they in fact, as I will proceed to demonstrate throughout this book, continue to animate many of the fundamental assumptions of new and emerging biometric systems.

Although such practices as physiognomics and phrenology are not included in traditional accounts of biometrics, I argue in this chapter that they must be seen as proto-biometric technologies, as they are fundamentally concerned with the measurement, calibration, classification and identification of bodily and behavioural features. Furthermore, both physiognomics and phrenology were premised on the notion that unique corporeal attributes could, in the hands of the skilled interpreter, reveal the inner "truths," including the moral and intellectual qualities, of the subject in question. Such a premise, as will be demonstrated later in this chapter, also underlay the development of inked fingerprint biometrics in the latter part of the nineteenth century. The proto-biometrics that I track in the course of this chapter effectively constituted the conditions of emergence for a technology such as inked fingerprints in the context of a specific historical juncture charged with a number of biopolitical concerns, including British colonial rule in nineteenth-century India and the need to develop a reliable system of classification and identification of suspect populations, and the subsequent transference of this biometric technology to a criminal-juridical context in metropolitan Europe.

Biometrics, in one form or another, has been deployed in human societies for millennia. Potters pressed their fingerprint into their finished work as signs of their individuating identities (Cole 2002, 60). Babylonian clay tablets dating back to 500 BCE evidence the manner in which business transactions were often sealed with the fingerprints of the concerned individuals (National Science and Technology Council [NSTC] 2006, 1). The systematisation of identificatory techniques based on corporeal attributes and signs, however, only begins to be consolidated in the context of early modern Europe. Valentin Groebner tracks the contemporary western preoccupation with identity and its corporeal signifiers back to this historical context. "Identity," he writes,

> is a medieval coinage. It was in common use in its Latin form *idemptitas* or *identitas* in medieval logic. Derived from *idem*, "the same," or *identidem*, "time and again," it denotes not uniqueness, but the features that the various elements of a group had in common.
>
> (Groebner 2007, 26)

Already encoded, etymologically, in the term "identity" is that seemingly contradictory logic that will proceed to haunt all biometric technologies based on individuating signatures and unique corporeal attributes; this contradictory or aporetic logic demands that individuating signatures must be iterable and reproducible, while also appearing unique in order to maintain their identificatory function.

Examining the early modern "prehistories" of systems of identification, Groebner (2007, 65) argues that "Modern identity papers can in fact be described as the combined outcome of those techniques developed between the thirteenth and sixteenth centuries, as watermarked, stamped, and signed papers bearing a seal and featuring a portrait." In his archaeology of identificatory systems based on locating unique corporeal attributes, Groebner (2007, 98) brings into focus the critical role that a subject's skin played in this process: "to medieval and early modern contemporaries ... what distinguished individuals from each other was located not inside the body, but on its outside, that is, in or on its skin ... Reading the skin—and writing on it—was an acknowledged technique for establishing the truth. A history of identification in Europe could begin literally on the skin." This early modern preoccupation with surfaces, as I will demonstrate later in this book, is precisely what gets overturned in the nineteenth century, as the "truth" of the body emerges as something that can be best accessed by plunging beyond corporeal surfaces and into the depths of the body.

The early modern processes of identification and documentation of body marks, including birthmarks, freckles, pockmarks and scars, became embedded within medico-legal frameworks that effectively transmuted individuating corporeal marks into "*corpus delicti*, or material evidence … At the end of the Middle Ages, skin marks became the legal paradigm of identification as embodied memory, as what was inscribed alike on the body and on paper" (Groebner 2007, 112). What Groebner establishes here is that foundational system of equivalence between individuating body marks, their textual transcription within official documents, and systems of identification that will underpin all biometrics, including the most cutting-edge digital technologies. Furthermore, the historical context of early modern Europe sees the development of systems of classification of bodily marks that are simultaneously invested with the ability to disclose a subject's *moral* character, thereby establishing an historical precedent for eighteenth-century physiognomics and nineteenth-century phrenology—with one critical difference: the individuating body marks of a subject were classified according to a subject's complexion, and not in relation to the distinguishing features of their face (physiognomics) or the topography of the brain (phrenology).

In tracing this taxonomic system of bodily marks, the colour of a subject's skin and the resultant moral characteristics that these signified, Groebner uncovers a buried history of the relation between skin colour and moral qualities that will eventually morph, as I discuss later, into raciological and racist theories instrumental in legitimating European colonial expansion. Indeed, in a pivotal moment in his thesis, Groebner (2007, 141) remarks on how this early modern European system of identification through bodily marks, complexion and related moral attributes was principally influenced by a focus on Europe's racialised and gendered others: "The history of personal description and its criteria, however, reveals that the history of self-description in Europe is effectively and fundamentally the history of the female Tartar slaves in fourteenth-century Florence, the millions of Canary Islanders and West Africans brought into servitude, and the exterminated inhabitants of the Caribbean islands." Operative here is the foundational role that *exteriority* plays in European identity formations, something Edward Said critically addressed in *Orientalism*. Europe, Said (1991, 20–21) argues, has unfailingly premised its multiple systems of self-identification on that othered figure or subject that must remain extraneous to it (Tartar slaves, for example), even as this same other is critically constitutive of European identity.

Groebner, in the course of his work, does not invoke Foucault's concept of the biopolitical, but, regardless, it is something that remains implicit in his archaeology of the historical conditions of emergence of identity papers in early modern Europe. In his discussion of the official implementation of identity papers, he brings into focus the manner in which European city authorities' concern to identify, register and monitor beggars and paupers "led to the creation of the identity paper in the literal sense" (Groebner 2007, 179).

Inscribed in this historical moment, and its textual instantiation of the identity paper, are a cluster of proto-biopolitical concerns that will continue to inform later identificatory systems. As I discuss later, such biopolitical concerns as the identification, registration, monitoring and governance of target groups and populations will be instrumental in the British development of inked fingerprinting in the context of nineteenth-century colonial India, and Alphonse Bertillon's development, in late nineteenth-century France, of his biometric system of *anthropometrical signalment*, with its focus on criminal populations.

Raciological Prehistories of Body Measurement

In the context of early modern Europe, as Groebner demonstrates, a discursive and iconographic process is set in train that is focused on mapping the body's features; the identificatory attributes of the body are then transposed to official identity papers. The systematisation of body measurement and modelling, however, only really began to consolidate itself as a scientific practice in the late fifteenth century in the context of the Renaissance. The anatomical mapping of the body was shaped by the intersection of the fields of medicine, mathematics and art. Perhaps the most famous exponent of this tradition is Leonardo da Vinci. His iconic "Vitruvian Man" depicts a Caucasian male encompassed within the geometric configurations of a circle and square. Da Vinci's drawing fundamentally established the template for this type of anatomical representation. Framed within these geometric coordinates, the Caucasian male emerges as the exemplar of perfect corporeal symmetry. A series of diagrammatic strategies function to constitute a visual regime that effectively regulates and disciplines representations of the human body. The da Vincian geometric grid establishes a type of matrix that disciplines the very contours of the human body and determines the normative dimensions and figurations of its features and its surfaces. Couched in Foucauldian terms, this geometric matrix, transposed to the body charts of medicine, becomes a technology of "body-power"; as such, it operates in the context of what Foucault (1982, 141) terms the "art of distributions," whereby a specific discipline assigns individuals particular subject positions and their attendant significations.

At the level of the visual image, the iconography of the normative or "ideal" body, and its predication on a geometrised system of measurements and calibrations, was imbricated early on with explicitly racialising agendas. The publication in 1791 of Petrus Camper's *A Treatise on the Natural Difference of Features in Persons of Different Countries and Periods of Life; and on Beauty, as Exhibited in Ancient Sculpture: With a New Method of Sketching Heads, National Features, and Portraits of Individuals, with Accuracy* enunciates the contours of such future disciplines as physical anthropology and anthropometry. In this text, Camper establishes a proto-evolutionary series of illustrations that pivot on measuring the facial angle of a subject. I say "proto-evolutionary" as one particular series of illustrations begins with the drawing of a tailed ape, encompasses the faces of an orang-utan, followed by the face of a young African male, a Calmuck (of the "Mongoloid race") and culminates in the Aryan face of a European male modelled on a classical sculpture of Apollo. Inscribed in this series is a racist teleology that would find its formal articulation in nineteenth-century social Darwinism. As James Elkins (1999, 170) observes, Charles Darwin "ascribes the beauty of Europeans in part to the powers of intellection: 'With civilized nations ... the increased size of the brain from greater intellectual activity, [has] produced a considerable effect on their general appearance when compared to savages'."

I draw particular attention to Camper's iconography of the human face as it established the racialising codes that proceeded to inform a number of aesthetic and scientific disciplines in both the eighteenth and nineteenth centuries. In his *Treatise*, Camper (1794, 1) developed a geometric formula for calculating the relation between facial angle and "the natural difference of features in persons of different countries." The racialising ideology that inflects Camper's (1794, 9) work is established early on in his thesis by his juxtaposition of an ape's skull with that of an African and a Calmuck:

> When in addition to the skull of a negro, I had procured one of a Calmuck, and had placed that of an ape contiguous to them both, I observed that a line, drawn along the forehead and the upper lip, indicated this difference in national physiognomy; and also pointed out the degree of similarity between a negro and the ape. By sketching some of these features upon a horizontal plane, I obtained the lines which mark the countenance, with their different angles. When I made these lines to incline forwards, I obtained the face of an antique; backwards, of a negro; still more backwards, the lines which mark an ape.

Already encoded in this series of cranial juxtapositions is a proto-evolutionary teleology whose dynamic hinges on articulating the difference between the civilised European and the ape-like African. In Camper's (1794, 84) schema, the antique refers to the absolute standard of beauty: the face of the Pythian Apollo, he writes, has "a proportion which in itself is constant with all our ideas of beauty." Camper acknowledges in his *Treatise* how he has been influenced by the German scholar Johann Joachim Winckelmann. He outlines how his work represents a labour to transmute Winckelmann's thesis on the absolute ideal of beauty, as established by the ancient Greeks, into a calculable formula "founded upon the rule of optics" (Camper 1794, 4). The difference between this ideal of European beauty and the African is, according to Camper's (1794, 42) calculations, quantifiable by 30 degrees: "The two extremities therefore of the facial line are from 70 to 100 degrees, from the negro to the Grecian antique; make it under 70, and you describe an ourang or an ape."

In his mapping of the intersection of aesthetics and theories of race, David Bindman (2002, 89) discusses the Eurocentric system of values that underpins Winckelmann's ideals of beauty (see also Potts 2000, 159–63). Winckelmann, he argues,

> did find deformity in what he believed to be salient bodily characteristics of non-European peoples. The horizontal eyes of the Chinese are "an offense against beauty," just as the squashed nose of Calmucks, Chinese and other peoples is "an irregularity," incompatible with the unity of form of the body. These "deformities" may well be caused by climatic conditions, as in the case of Africans: "The mouth swollen and raised, such as the Negroes have in common with the monkeys of their country, is a superfluous excrescence, a swelling caused by the heat of the climate."

This history of body measurement, its racial inflections and the construction of what I will term the *template white body* are not unconnected to contemporary biometrics. The scripting of non-white phenotypical features, such as "horizontal eyes," "squashed noses," and "swollen mouths" as "deformities" and "superfluous excrescences," results in their being finessed away from the schematic white bodies that will supply the templates for the calibration of body measurements in the future. As I demonstrate in Chapter 2, black skin or "Asian" fingerprints, as "superfluous excrescences" to the template white body, often cannot be read by some biometric systems whose image acquisition systems are predicated on the template white body.

Uncoincidentally, as Martin Bernal has argued in *Black Athena*, Winckelmann was one of the key figures in the establishment of what he terms the "Aryan model" of history. Although he does not mention Camper in his text, Bernal maps in detail the key European thinkers of the mid-to late eighteenth century who were instrumental in developing racial theories of white supremacy. In the historical context in which Camper was writing, Bernal (1987, 217, 219) mentions the historian and anthropologist Christoph Meiners ("later to be honoured by the Nazis as a founder of racial theory"), who "wrote 'progressive' Romantic histories of peoples whom he divided categorically into white, courageous, free, etc., and the black, ugly, etc. The spectrum ranged from chimpanzees through Hottentots and others to Germans and Celts." The other key figure within this historical context is Johann F. Blumenbach, professor of natural history at Göttingen University, who first theorised the concept of the Caucasian in 1795: "According to him the white or Caucasian was the first and the most beautiful and talented race, from which all other races had degenerated to become Chinese, Negroes, etc" (Bernal 1987, 219). Blumenbach was instrumental in developing the theory of chromometrics, designating five different races to which he allocated a specific colour: "Caucasian or white, Mongolian or yellow, Ethiopian or black,

American or red, and Malayan or brown" (de Waal Malefijt 1974, 259). Blumenbach's chromometric schema, in its mixing of science, race and colour, effectively operated as a form of "racial geometry" (Gould 1996, 401) within which the colour of a subject's skin determined her or his allocation along a racial hierarchy governed by whiteness.

I briefly sketch the historical context within which Camper developed his theories in order to problematise Stephen Jay Gould's (1987, 15) assertion that "Camper did not define the facial angle as a device for ranking races or nations by innate worth or intellect ... Camper did not locate his immediate motive for the facial angle in descriptive anthropology of actual humans, but in a much loftier problem—no less than the definition of beauty itself." I have no interest here in re-animating the intentionalist fallacy by attempting to discover the "true intentionality" of Camper's thesis on facial angle. Rather, my interest is in locating Camper's *Treatise* within a specific historical field in order to elucidate the discursive effects of his work. As such, "the much loftier problem—no less than the definition of beauty itself" produces a series of racialising effects that refuse to be neutralised by the apparently transcendent category of pure aesthetics. Indeed, as John Jackson and Nadine Weidman (2006, 42) argue, Camper was a formative influence in shaping Georges Cuvier's racist hierarchy:

> He ordered the animal and human races along a graded scale of intelligence based on their facial angle, an idea he borrowed from Camper and made more sophisticated by his own new comparative anatomical measurements and methods. By correlating facial angle and cranial measurements with perceived mental and moral qualities, Cuvier believed he had proved that the Ethiopian race was at the bottom of the scale, closest to apes, and that its condition was foreordained and unchangeable.

Operative in Cuvier's racial hierarchy is that inaugural moment of what Denise Ferreira da Silva (2007, 129) terms "the manufacturing of the analytics of raciality." Silva (2007, 129), in her painstaking mapping of raciality as a form of globality that has been instrumental in the production of asymmetrical subject positions and relations of power, identifies this analytics of race as:

> the political-symbolic arsenal that transformed the categories employed by eighteenth-century naturalists and philosophers into scientific signifiers that produced the human body, social configurations, and minds as effects of the tools productive of *nomos*. That is, the signifying strategies of *scientia racialis* produced a strategy of engulfment, the racial, as a signifier of human difference that presupposed that scientific reason accounts for the various existing modes of being human.

As Silva proceeds to demonstrate, this political-symbolic arsenal of race as globality continues to ramify across both ontological and epistemological levels.

The complex imbrication of science with moral categories of value and aesthetics has also been brought into focus by Ludmilla Jordanova (1993, 123–133) who has demonstrated, in her work on physiognomic texts published in eighteenth-century Europe, how physiognomy was "at once a moral and aesthetic language—appearance reveals good and bad characters through beauty and ugliness." In his *The Mismeasure of Man*, Gould (1996, 66) delineates a thesis on racial classification that clearly resonates with this position: "In the first formal definition of human races in modern taxonomic terms, Linneaus mixed character with anatomy (*Systema naturae*, 1758). *Homo sapiens afer* (the African black), he proclaimed, is 'ruled by caprice'; *Homo sapiens europeaus* is 'ruled by customs'."

In her analysis of eighteenth-century pathognomics (as the study of the features that mark the face in the course of expressing emotions), Barbara Maria Stafford (1997, 128) illustrates how the profile of the African black "was taken to signify the ideal of stupidity and recalcitrance." Focusing on the deployment of Camper's *Treatise* within a range of disciplines, Stafford (1997, 115) argues that his facial profiles "could be understood as constituting an infallible anthropological method for visibilizing psychic perfection and imperfection in different ethnic groups." In other words, what is being articulated here, in historical and theoretical terms, is the absolute impossibility of disarticulating theories of beauty from ideologically loaded concepts of race. In tracing this genealogy of body measurement and its relation to race via the work of Camper, I am attempting to map what George Mosse (1978, 2–3) identifies as "a cardinal feature of modern racism": the "continuous transition from science to aesthetics": "Human nature came to be defined in aesthetic terms, with significant stress on the outward physical signs of inner rationality and harmony."

Camper's aesthetics are founded on the racialising principle that European bodies are the normative standard, the templates, by which beauty can be measured. "The Calmuck," writes Camper (1794, 18),

> compared with ourselves, and more particularly with the most celebrated figures of antiquity, are deemed the ugliest of all inhabitants of the earth. Their faces are flat, and very broad from one cheek-bone to the other; the nose is so flat, that the sight penetrates into the nostrils; the lips are thick, and the uppermost lip is long.

The corporeal geometries of proportion, symmetry and balance that constitute his aesthetic are predicated on a Eurocentric norm: "When the face is seen in profile, the breadth ough [sic] not to exceed the height, as in the negro and Calmuck: in us they are nearly equal" (Camper 1794, 90). The template white body hinges on an unstated but symbolically violent operation of negation and exclusion. The template white body can only signify its typicality, its exemplarity and its schematic universalism through the segregation of non-white phenotypical features that, tautologically, code for race. In terms of the normative frames of white supremacism and its regimes of representation, to qualify "race" as that which is "non-white" is already a redundant move. These racial differences are extrinsic to the ideal human form (that is, the white template body); indeed, they are racial precisely because they are extrinsic, that is, *other*. As such, non-white phenotypical features function to contaminate and compromise the normative and universal status of the (non-raced) white body with what Winckelmann, as mentioned earlier, termed "superfluous excrescences."

In the U.S. context, Camper's racial profiling sequence, beginning with the ape and followed by the consecutive juxtaposition of an African male and the head of Apollo, was, during the antebellum period, effectively transposed to the field of physical anthropology by such slavery apologetics as the scientist, George Giddon, and the physician, Josiah Nott. William Coleman (1971, 98–99), in his *Biology in the Nineteenth Century*, demonstrates how:

> Josiah Nott verbally and visually urged, on ostensibly scientific grounds, the near-bestial condition of a distinct Negro "race" of mankind: "a man must be blind not to be struck by similitudes between some of the lower races of mankind, viewed as connecting links in the animal kingdom; nor can it be rationally affirmed, that the Orang-Outan and Chimpanzee are more widely separated from certain African and Oceanic Negroes than the latter from the Teutonic or Pelasgic types."

The Traffic in Bodies: The Iconographic Data of Non-White Bodies and the Ontological/Epistemological Split

Even as the European iconographic tradition of anatomical modelling and mapping, in the fields of science and medicine, excluded non-white bodies from the symbolic level of representation, they in fact played fundamental roles in supplying the "data" or corpo-ontological infrastructure that was later transmuted into higher-order western medico-legal knowledge. In order to begin to address this ontological/epistemological split that I have just outlined, I want to return to the work of Petrus Camper. Reading Camper, I was struck by three casual, passing references in his text. Let me reproduce these three fragments: "When in addition to the skull of a negro, I had procured one of a Calmuck and had placed that of an ape contiguous to them both" (1794, 9); Camper (1794, 22) then describes the dissection of a foetus, "which was about six months old," of a "female negro"; finally, Camper (1794, 23) writes:

> In the year 1758, I dissected publicly at the anatomical theatre at Amsterdam, the body of a negro lad, about eleven years of age.

> This afforded me an opportunity of demonstrating all those diversities in the cranium, which nature had effectuated.

These three passing references to non-white bodies in Camper's text are not insignificant. Inscribed in the margins of these casual references is an entire history in the traffic of non-white bodies, as bodies productively put to service the epistemic economies of western medicine and science. The term "procured" begs the question: how were these bodies made available as so much merchandise to be purchased on the market? The references to these non-white bodies disclose what remains unspeakable yet silently and insistently informs Camper's *Treatise*: the operations of colonialism and economies of slavery within the Dutch empire of the eighteenth century. It is in the context of the interlacing economies of slavery, colonialism and western science, that the "procurement" of the body of a "female negro" or a Calmuck becomes intelligible. From the sixteenth century until well into the eighteenth century, Pieter Geyl (1968, 8–3) notes, the "The Dutch took a very active part" in the trade of slaves; the slaves "were in various places bought by the thousand every year ... particularly on the Gold Coast [of Africa] and the so-called 'Bucht' (Bend)." "The abominations inherent in the whole business," writes Geyl (1968, 83), "were regarded with composure, because heathens were not looked upon as fellow-men." The Dutch trade in slaves is located, indeed, within the very salubrious confines of the capital, Amsterdam, where the Dutch entrepreneur, Balthasar Coymans, had a major contract in the transport of slaves from 1685 to 1689 (Geyl 1968, 372).

Placed in the context of the anatomical theatre of Amsterdam, where Camper performed his public autopsies, the non-white body is transposed from its colonial marginality to the commanding position of centre-stage. From this locus, the non-white body becomes legible as spectacle, as the corpse is made to offer itself up to the inquiring gaze of the scientific community. During the process of the public post-mortem, the non-white body, as embodied metaphor for the Dark Continent, is penetrated and literally turned inside out in an analogue of the colonial voyage of discovery. The value of the non-white body within these western epistemic economies is, however, clearly circumscribed. Non-white bodies have supplied the material/corporeal substratum that has been assiduously mined by western science. As another variation on the colonial role of native informant, the non-white body has supplied the lumpen data that has then been converted into higher-order knowledge by the western scientist. These non-white bodies have supplied the critical knowledge of the human body that enabled the production of the corporeal cartographies and atlases used in medical schools, academies, forensic laboratories and so on.

In the context of the U.S., Michael Sappol (2002, 17), in his *A Traffic of Dead Bodies*, has mapped the repeated robbing of African American bodies by white anatomy students from graves in the Negroes Burial Ground in

New York City during the eighteenth century, despite protestations from the black community to have this practice outlawed. Sappol (2002, 258) writes how "blacks involuntarily supplied a disproportionate number of subjects for medical school dissecting tables." In the Australian context, the bodies of Indigenous peoples were illegally "procured" after acts of violent expropriation, including massacres and the robbing of graves. The Aboriginal writer, Kevin Gilbert (1988, xx), bears testimony to this violent history:

> In my country, Wiradjuri, a large mob of my countrymen, women and children were herded and driven like sheep before the guns [of English soldiers] to the big swamps near Bathurst. There they were "dispersed" with guns and clubs, whereupon these pioneering, head-hunting whites cut off a large number of the people's heads, boiled them down in buckets and sent 45 of the skulls and other bones off to Britain. In much the same way, they took Pemulwy's [Aboriginal warrior] head, pickled it and sent it off to the Joseph Banks collection in England.

Even as they underpin the very production of western corporeal cartographies and epistemologies, non-white bodies become precisely that which cannot be represented as such within the visual economies of elite western disciplines such as medicine and science (Pugliese 2005). This is exemplified by the manner in which the non-white body, within the specific field of western anatomical atlases, has been effaced from the level of symbolic representation. Non-white bodies, once situated within this medico-scientific context, evidence, then, the sort of ontological/epistemological split "that pits subaltern being against elite knowing" (Spivak 1987, 268). This structural asymmetry ensures the literal unrepresentability of the non-white body in anatomical atlases of western medicine, except as examples of gross anomaly (for example, Saartjie Baartman, the so-called "Hottentot Venus," paraded across Europe in order to exhibit her protruding buttocks and genitalia [Gilman 1986]). Even as they underpin the production of specific epistemologies, non-white bodies must remain what Gayatri Spivak (1987, 204), citing Derrida, terms "the blank part of the text." Even as they supply the corporeal materiality of flesh—as the matter in hand that is dissected and analysed—non-white bodies must needs undergo a type of "decorporealized etherealism" (Barker 1984, 94), whereby they literally and symbolically disappear in the process of being converted into schematised white bodies of knowledge.

The Geometry of the Face: Physiognomics and Facial Indices of Character

Although usually seen as emerging as a defined discipline in eighteenth-century Europe, physiognomics has a long and complex history. Physiognomics is predicated on reading the facial features, or physiognomy, of a subject in order to determine his or her character. Variant forms of physiognomy can be traced back to ancient Greece, where Physionymas is identified as the founder of the discipline (Groebner 2007, 119), and even further back to paleobabylonian culture in ancient Mesopotamia (Rivers 1994, 18). In his tracking of the identificatory dimensions of bodily signs in early modern Europe, Groebner (2007, 120) argues that a form of physiognomy was practised in which "nature" was seen to determine "a person's outward appearance, constituting an individual whose character can be read from the exterior." The key corporeal index for the physiognomic evaluation of a person's character was complexion. Groebner unpacks a complex chromatics of complexion that is inscribed with a system of symbolic indices that enable the trained physiognomist to read for a subject's inner state and character. He names the trained physiognomist an "expert narrator" "who defined the allegedly intrinsic nature of human individuals through the *signa* permanently inscribed on their epidermis" (Groebner 2007, 145). Already encoded in these early forms of physiognomics are three fundamental presuppositions: that external corporeal signs are indices of inner, moral and intellectual, qualities; that the relation between the two is "natural" and innate; and that the decoding of these corporeal indices requires an expert hermeneut who can interpret bodily signs or *signa*. As I will discuss in some detail in Chapter Five, these physiognomic presuppositions continue to inform contemporary reincarnations of the discipline in the context of such technologies as Brain Fingerprinting, with its promise that it can read cerebral activity in order to disclose truth and lies.

Even though physiognomics has a long and complex history, as a discipline it really only achieved high status and fame once it was systematised ("as a science of determining human character traits and tendencies from physical appearance" [Rivers 1994, 69]) in the latter years of the eighteenth century by the Swiss physiognomist, Johann Kaspar Lavater (1741–1801). The publication of Lavater's *Physiognomische Fragmente zur Beförderyrung der Menschenerkenntis und Menschenliebe* (1775–78) was a landmark event in European culture. It was subsequently translated into a number of European languages and it established itself as the canonical work in the discipline. Lavater's work presents the reader with a vast catalogue of illustrations that represent the variety of human physiognomics (from profiles to full-frontal views encompassing an array of different expressions and emotional states), while also outlining the interpretative methods by which to decode facial features in order to disclose the inner qualities of the subject in question. As with so many practices founded on scientific principles (regardless of their historical mutability), Lavater promised to deliver in his systematisation of physiognomics a series of immutable rules that would guide the physiognomist in her or his hermeneutic practice. As Christopher Rivers (1994, 65) observes, "Lavater's work is predicated on a desire to reinscribe physiognomy as a discourse of the absolute, and to establish thereby (once and for all) a physiognomical 'alphabet' which transcends all human will, deception, and ambiguity." As I discuss in Chapter Five, this scientific promise to deliver a method that will "transcend human will, deception, and ambiguity" is what continues to inform such technologies as Brain Fingerprinting, with its promise that it can circumvent the will and deceptive strategies of the subject undergoing, for example, an electroencephalographic (EEG) scan in order to reveal whether or not they are telling lies. Furthermore, physiognomics can be seen to be a formative precursor of a contemporary technology such as Brain Fingerprinting because of its linking of "medical diagnostics to textual criticism [as a form of hermeneutics] and cerebral expertise" (Stafford 1997, 84).

Underpinning Lavater's scientific systematisation of physiognomics, with its vast classificatory distribution of faces and their attendant expressions and psychic and emotional states, is a normative disciplinarity. The range of physiognomies can only be decoded in terms of their moral qualities when measured and evaluated against a series of normative assumptions. The "crux of physiognomical practice," Lucy Hartley (2001, 2) underscores, "is a classificatory act which functions in a profoundly normative manner in so far as it takes a particular expression as the exemplification of a general kind and then uses this to describe the character of the individual." Although physiognomics had its heyday in the latter half of the eighteenth century, its influence continued to be felt well into the nineteenth century, where it informed a number of cultural and scientific practices (Hartley 2001). It was, however, as a scientific practice that promised to reveal the inner qualities of the human subject by reading corporeal signs, largely supplanted by the nineteenth-century discipline of phrenology. Phrenology turned its critical attention from the face to the brain, promising, as physiognomics had done, to be able to reveal the true, innate and essential qualities of the subject in question through a reading of the topography of the brain.

Phrenology: Brain Locationism, Cranial Bumps and the Topography of Character

Phrenology was premised on the belief that the scientific study of the topography of a subject's brain, and her or his cranial bumps and indentations, would reveal the intellectual, affective and moral qualities of the person under examination. Phrenology was developed in the late eighteenth century by the Austrian physician Franz Joseph Gall (1758–1828). It was Gall's assistant, however, Johann Gaspar Spurzheim (1776–1832) who popularised Gall's theory, embarking on extensive lecture tours across Europe, the U. K. and the U. S. It was Spurzheim who developed the now iconic image of the phrenological head chart, with its numbered divisions of the various regions of the brain and their attendant moral, affective and intellectual attributes. While on one of his lecture tours, Spurzheim met the lawyer George Combe

(1788–1858) in Edinburgh. Combe was quickly converted to phrenology, becoming its leading exponent in the U.K. and publishing a number of influential books on the subject. In his *Elements of Phrenology* (1834), Combe (and Spurzheim 2007 [1834], 45) declared phrenology "a new science" that would revolutionise studies of the brain and mind. John van Wyhe (2004, 1), in *Phrenology and the Origins of Victorian Scientific Naturalism*, writes that: "The science of phrenology, and the writings of George Combe in particular, had profound effects on Victorian culture. The historian Roger Cooter compared this impact to 'Newtonianism in the eighteenth century or structuralism in the twentieth, one of those powerful but largely unseen influences on modern thought'."

Phrenology can be seen to be the successor science to physiognomics. Transposing physiognomics' reliance on the reading of the structural features of the face, as a way of discerning the mental and moral qualities of a subject, to the brain and the cranium, phrenology proceeded to map these inner qualities on a localised terrain. The brain was segmented into various regions or "organs," with each cerebral locality assigned a specific affective, intellectual or moral attribute; so, for example, the "organ" of "veneration" was seen to be "situated at the middle of the coronal aspect of the brain, at the bregma or fontanel of anatomists" (Combe 2008 [1834], 58–59). The intersection of physiognomics with phrenology is made clear in van Wyhe's (2004, 15) description of Gall's development of a theory of cerebral localisation based on a "diagnostic cum character divination method via physiognomical head reading." As with physiognomics, phrenology claimed to offer an unmediated view into the inner workings of the subject under examination, even as both practices were fundamentally reliant on the mediative role of interpretation. This paradox is clearly evidenced in Combe and Spurzheim's (2007 [1834], 119) declaration that "Nature and positive facts are the only authority which Phrenologists acknowledge," while simultaneously describing phrenology as "the science [that] is held to be *a true interpretation of nature*" (Combe 2008 [1834], 139). That the positivity of "natural facts" can only come into epistemological being through the instrumental role of hermeneutic practice is what must be forgotten in order for a particular practice to maintain its claim to scientificity, and its attendant attributes of objectivity and unmediated impartiality. As I argue throughout the course of this book, this crucial movement of occlusion continues to inscribe many of the contemporary claims made with regard to the objective and unmediated status of biometric technologies.

Phrenology's belief "that irrefragable Nature is truth" (van Wyhe 2004, 50) enabled the establishment of a science that proffered normative and moral evaluations of humans based on "natural" evidence. Drawing on the racial theories developed by the likes of Blumenbach and Cuvier, phrenology began to systematise its empiricist claims regarding the relation between the brain and the specific qualities of mind along well-established racial and gendered hierarchies: "*Man* was arranged in a hierarchical scale of superiority and inferiority. The scale began with non-European races, especially those with dark skins 'whose brains are inferior' at the bottom, and western Europeans, like himself [Combe], at the top" (van Wyhe 2004, 120). In his *Elements of Phrenology*, Combe delineates a neuro-topography of intellectual and moral attributes that effectively replicates nineteenth-century colonial cartography and its racialised evaluation of particular non-western populations according to their geographical location. For example, in Indigenous Australians, "New Hollanders," the "organ of causality" is described as being "very deficient" (Combe 2008 [1834], 97). Combe (2008 [1834], 24) illustrates this self-evident empirical assessment by juxtaposing the "small head of an idiot" with that of a "New Hollander." This "scientific" assessment of Indigenous Australians perfectly dovetailed with the view of the British colonial invaders and administrators of the continent, who described Aboriginals as "'frozen moments' from the Palaeolithic past," "the most disgusting type of living savages" (cited in Russell 2001, 13).

Phrenology's neural topography/colonial cartography pivots on the division between subjects that are quintessentially human (Europeans) and those non-western subaltern populations who are, once examined under the objectifying lens of western science, discovered to be "naturally" closer to animals. So, for example, faced with the conundrum that "The brains of Charibs seem to be equal in absolute size to the those of Europeans," Combe (2008, 122–23) proposes that "the chief development of the former is in the animal organs, of the latter in the organs of sentiment and intellect; and no Phrenologist would expect the one to be equal in intelligence and morality to the other, merely because their brains are equal in absolute magnitude." Phrenology's comparative method of brain and skull analysis ensured the discipline's participation in the European traffic in Indigenous bodies that, as I outlined earlier, has been constitutive of Western anatomo-scientific knowledge.

Such scientific racial classifications and racist evaluations of colonised populations effectively functioned to legitimate colonial occupation and the acts of colonial genocide perpetrated by the colonisers against Indigenous peoples. A. Dirk Moses has traced the close imbrication of scientific racism and colonial genocide. He cites, amongst others, the British novelist, Anthony Trollope, observing during his travels in Australia that "The Aborigines were 'ineradicably savage' … the male possessed the deportment 'of a sapient monkey' … as well as suffering from a 'low physiognomy' that rendered him lazy and useless. 'It is their fate to be abolished'" (Moses 2005, 5). Situated within this context of scientific racism and colonialism, phrenology can be seen to be implicated in the biopolitical programs administered within nineteenth-century Europe's colonies.

Operative in phrenology's scientific articulation of the relation between race, and the degree of one's human status or worth is what Foucault (2003, 60) identifies as a "materialist anatomo-physiology" that will proceed to base its biologically racist claims on the self-evidence of nature and the body. The Italian criminal anthropologist Cesare Lombroso (1835–1909) built on phrenology's positivist anatomo-physiology—and its claim that intellectual, moral and affective qualities could be identified through their topographic delineation of cerebral regions—in the development of his theory of the "born criminal." The conceptual apparatus of Lombroso's theory of the born criminal, as outlined in *Criminal Man*, is premised on a series of biopolitical assumptions inscribed with normative and Eurocentric notions of race, gender, (dis)ability and sexuality. Criminals, in Lombroso's positivist schema, are identified and positioned in relation to their degree of deviation from normative standards. Lombroso's (2007 [1876–97], 48–49) exhaustive cataloguing of criminal indices always begins with the tracing of what he terms "rates of abnormality," which are concerned with "congenital differences" that deviate from his presupposed norm: "Criminals have the following rates of abnormality: 61 percent exhibit fusion of the cranial bones; 92 percent, prognathism or an ape-like thrust of the lower face." "These features," Lombroso (2007 [1876–97], 49) concludes, "recall the black American and Mongol races and, above all, prehistoric man much more than the white races." Incorporating the caucacentric racial hierarchies established in the eighteenth century, the "savage" is once again invoked as the ancestral relation of the criminal: "The study of cranial anomalies suggest that the criminal is closer to the savage than to the madman" (Lombroso 2007 [1876–97], 364). In Lombroso's work, the "savage" or "primitive type" informs both male and female "born criminals"; the female criminal, however, is positioned as even further down the evolutionary scale than the male: "atavistically," the female criminal "is nearer to her primitive origin than the male" (Lombroso and Ferrero 2004 [1893], 146). This also accounts for one of the defining features of the female criminal: her "virility." "To understand the significance and atavistic origin of virility, we have to keep in mind that it is one of the outstanding traits of savage woman. We can see this best by looking at portraits of American Indian and Negro Venuses … It is difficult to believe that these are really women, so huge are their jaws and cheekbones, so hard and thick their features" (Lombroso and Ferrero 2004 [1893], 149–50). Lombroso's gallery of racialised and gendered criminal types was illustrated by hundreds of photographs and drawings that visually evidenced each type of criminal. Operative in Lombroso's work is a visual regime that works self-evidently to capture the identificatory features of a particular criminal type and that lays the discursive foundations for the emergence of the police mug shot and contemporary technologies of preventative criminology based on visual regimes of "preincident indices" (see Chapter 3).

Lombroso's gallery of criminal types was further enhanced by Francis Galton (1822–1911), who is credited with the development of eugenics and the systematisation of inked fingerprints into a classificatory system that proved invaluable to the practice of criminal identification (Cole 2002, 74–81). Galton's iconography of criminal types was established through his superimposition of "multiple exposures of a group of criminals onto a single photographic plate. The resulting ghostly 'composite' image revealed the physiognomic attributes common to a set of criminals, thus allowing authorities and researchers to 'see' the criminal type" (Cole 2002, 24). Galton's reliance on the face of the photographed subject to offer the visually identifiable features of criminality, through a type of composite distillation, discloses his reliance on a form of biological essentialism that self-evidently discloses innate moral qualities. Viewed in terms of regimes of visuality, as visual regimes of power/knowledge that mediate and regulate the physiological process of seeing, what is operative in Galton's images of the faces of criminal types is what I want to term the *faciality of the face*.

The faciality of the face is not a tautology; rather, it brings into focus the inscriptive schemata and discourses that render a face culturally intelligible precisely as *identifiable face* in terms of dominant discourses, tacit knowledges and assumptions. There is never any unmediated visual encounter with the face; rather, the moment of visual apprehension and comprehension is always already marked by an inscriptive cultural and discursive schematicity of the face before one's gaze: it is the schematicity of this faciality that renders the face culturally intelligible and identifiable as *face*. In the photographs that illustrate Galton's gallery of criminal types, the signed exteriority of the criminal face, branded with the imprimatur that identifies the criminal type (for example, "hotel thieves," "pickpockets," "burglars" and so on) enunciates the incarnation of a criminal interiority: the distilled or composite phenotypicality of the subject in question—low forehead, swarthy skin, dark eyes—appears as a schematised index of his or her criminal nature. As David Cole (2002, 26) observes, "These experiments in composite photography, however, never succeeded in discovering any unequivocal physiognomic marker of the criminal type. Instead, criminal physiognomy became a crude pretext for prejudices, such as the association of criminality with 'swarthy' Eastern and Southern Europeans, Jews, Gypsies and others who simply looked lower class."

In his mapping of what he termed "the organic roots of crime" (2007 [1876–97], 338), Lombroso drew upon physiognomic typologies in order to construct his profiles of born criminals: "each type of crime," he writes, "is committed by men with particular physiognomic characteristics" (2007 [1876–97], 51). "The ability to ascertain moral temperament from physiognomy and the cranium," he argues, "is a scientific method of discerning character" (2007 [1876–97], 233). Criminal anthropology's use of both physiognomy and phrenology underscores the manner in which the measurement and classification of corporeal features were instrumental in producing the identifiable biotypologies of the "born criminal." As David Horn (2003, 6) notes, "The production of 'the criminal'—and, more specifically, the criminal body—as objects of knowledge in late-nineteenth-century human sciences was dependent on technologies of counting and calculation that had worked to reconfigure crime as a social problem."

Anthropometry: The Biopolitics of Measuring Bodies

A survey of the scientific field in nineteenth-century Europe reveals the proliferation of disciplines conceptually premised on the measurement of bodies. Encompassed under the superordinate term *anthropometry* (the "systematized judicious observing and measuring" of the body [Hrdlicka 1939, 3]) is a series of disciplinary subdivisions that indicate the extent of the field and its reach across and into every external and internal area of the body: craniometry (measurement of the cranium); osteometry (measurement of bones); organometry (measurement of organs); encephalometry (measurement of the brain); psychometry (measurement of mental functions); physiometry (measurement of bodily functions); somatometry (measurement of the living body); and biometry (the application of mathematics to biology) (Hrdlicka 1939, 3). Anthropometry, in all of its polymorphous forms, exemplifies the manner in which, in the course of the nineteenth century, two technologies of the body—disciplinary and biopolitical—became intertwined. Anthropometry is, on the one hand, focused on disciplining the body in question through the deployment of a complex repertoire of technologies and techniques of measurement, classification and hierarchisation, while, on the other hand, it proceeds to draw up "statistical estimates" and "overall measures" in order to "intervene at the level at which these general phenomena are determined, to intervene at the level of their generality" (Foucault 2003, 246).

In Ales Hrdlicka's (1939, 10) canonical treatise, *Practical Anthropometry*, the discipline is shown to have a long history that can be traced back to ancient civilisations concerned "in the formulating of various standards and 'canons' for the human body; but above all in the XIX century, with studies on the human races, on the skeletal remains of early man, and in those of human growth." The studies on human races, and the discursively related field of the study of the skeletal remains of early man, are identified as two of the cornerstone practices instrumental in the modern development of the discipline. As I discussed earlier, the body of the non-western other has played a foundational ontological role in the development of western medical and scientific epistemologies. And, in the context of the nineteenth century, the body of the non-western subject becomes enmeshed within biopolitical programs of measurement, enumeration, and statistical deviation against normative standards. These biopolitical programs are primarily driven by concerns to regulate, administer and hierarchise (Foucault 1990, 144), and they are instrumental in the implementation of colonial rule. Hrdlicka (1939, 39), for example, discusses the uses to which he has put practical anthropometry in the context of his anthropological work:

> Among the American Indians, a reliable and valuable test for the determination of admixture with Whites has been found by the author during the special task of identifying, for the U.S. Department of Justice (1915–17), remaining fullbloods in the large tribe of the Ojibway.

Working under the auspices of the U.S. government, Hrdlicka's (1939, 39–40) test for the anthropometric identification and eventual segregation of "fullblood" American Indians from "mixed bloods":

> consists in exposing the chest to about the middle of the sternum in men, to the upper limits of the breasts in women, in drawing with the thumb nail of the observer, with some force but not enough to hurt the subject, vertical lines, from the upper part of the chest downward, in the middle of each sternum. If the subject is a fullblood, the lines made will remain faint or but moderately marked, with a narrow and soon evanescing darkening or duskiness along each line. If there is any White blood in the individual the skin reaction will be more marked, the lines will show broader, be reddish to red, and plainly more durable. And the more White admixture the more pronounced will be all these features.

Operative here is a test that both *produces race,* through the invasive dermographic inscription of the anthropologist's thumbnail across the body of the subject, and *racialised subjects*, through the resultant identification of the subject's racial "blood quantum." Race is here scientifically produced through the deployment of an apparatus of knowledge (anthropometry); through the mobilisation of an observational method (the clinical gaze, calculated to remain neutral and scientifically detached in the face of the objectified subject of inquiry); through the deployment of a disciplinary code of normativity (which sets the colour-gauge for the normative reaction of skin that has been inscribed by the anthropologist's thumbnail); and, finally, through the application of techniques of verification (the visible chromatic effects of a dermographic inscription, once measured against the normative colour-gauge, will disclose hidden "blood quanta"). The question of race is here literally and symbolically resolved through a form of writing on the body with the stylus of the anthropologist's nail, producing an intextuated body, literally a racial dermography that can only be decoded by the scientist. Hrdlicka's (1939, 39) test is premised on the notion of "markedly differing circulatory or blood reaction in the fullbloods and in those with white admixture." "White" blood, in this biopolitical schema, will tell. Whiteness, in keeping with its position at the top of the evolutionary hierarchy, is invested with a more sensitised quality ("the skin reaction will be more durable") against the more muted and dusky effects of the non-white "fullbloods." As such, as Hrdlicka (1939, 39) concludes, "this test is both easy and decisive, for legal as well as scientific purposes."

There is more at stake, however, in Hrdlicka's test than the production of race and racialised subjects. As absurd as Hrdlicka's test might appear in a contemporary light, it is in fact fundamentally constituted by an historically contingent scientific logic or rationality, what Foucault (2003, 55) would term the "rationality of technical procedure." The rationality of technical procedure that underpins this test also invests it with the politico-juridical status that enables it to be put to work on behalf of the colonial state, the U.S. Department of Justice in particular, and its biopolitical laws. This anthropometric test must be seen as a scientifico-legal technique of domination producing instrumentalised subjects that are, in turn, subjugated to the biopolitical laws of the state. Situated within the U.S. history of biopolitical regimes of racial segregation based on "blood quantum" lines, the "legal purposes" to which Hrdlicka alludes, refer to the U.S. government's administration of Native American lives and land under the *General Allotment Act* or *Dawes Act* (1887). The *General Allotment Act* formally legislated Native American identity on the grounds of biological racism through its construct of "racial blood." The separation of "fullblood" from "mixed blood" Native Americans that this Act enabled was used in order to determine the allotment of tribal lands according to "blood quantum standards." As Paul Taylor (2008, 146) notes, the "blood quantum" segregation of Native Americans played a crucial role in the colonial expropriation of Indian land: "Since whatever land was left over could be sold, there were substantial economic interests to be served by finding as few ['fullblood'] Indians as possible."

Hrdlicka's test for "blood quanta" had its equivalents in the Australian colonial context, where white anthropologists worked in the service of the biopolitical state in order to identify and segregate "mixed blood" Indigenous children from "full bloods." Working in the field, and deploying another set of anthropometric technologies, including colour filters that, once placed against the skin of the subject, scientifically determined their "blood quantum" status, Australian anthropologists were instrumental in the state production of colonial assimilation and the resultant Stolen Generations (see Report of the National Inquiry into the Separation of Aboriginal and Torres Strait Islander Children from Their Families 1997). Aboriginal and Torres Strait Islander children of mixed parentage were, under this biopolitical program, forcibly removed from their parents and placed either in State-run homes or farmed out in conditions of servitude to white domestic households or pastoral stations. Once again, one of key colonial effects of this devastating biopolitical program was to reduce the number of Indigenous claimants to their expropriated land.

Underpinning scientific promotions of the various forms of anthropometry, including contemporary biometrics, as I will discuss later, is the belief that such practices are, in Hrdlicka's (1939, 12) words, distinguished by "the complete elimination of personal bias." This is a claim that continues to be made in contemporary biometric literature. The Eurocentric system of values and the legal and political investments that underpin such scientific practices must be effaced in order to construct a view of science that transcends the very socio-political conditions that at once constitute it and enable its operations. The *General Allotment Act*, its "blood quantum" hierarchies and standards, and the implementation of this legislation by the anthropologist's application of practical anthropometry—all effectively evidence the operations of biopower. Principally concerned with the body and its relation to structures of power, "Bio-power brought life and its mechanisms into the realm of explicit calculations and made knowledge/power an agent of transformation of human life" (Foucault 1990, 143). Biopower effectively colonises the body, overlaying it with calculatory grids and geometrically inscribing it with formulae that will transform it into an object of knowledge and power. Furthermore, the separation of Native Americans and Indigenous Australians along racialised "blood quantum" lines also evidences the manner in which biopower actually constitutes bodies and subjects, constructing such categories as "fullblood" or "mixed blood"; such categories must be viewed as corporeal realities that produce legal, political, economic and social effects.

At virtually every level, race, and its varied intersections with the categories of gender, sexuality, (dis)ability and class, plays a foundational point of reference in defining, through anthropometric apparatuses, the self-identity of the European norm against all its deviations, degenerations and anomalies. Race is one of the foundational categories of the European archive, insistently establishing the conditions of possibility for the vast range of its epistemologies, including philosophy, science, law, anthropology, aesthetics and so on. It is not a case, however, of this foundational category operating in a seamless, transhistorical manner; rather, race is transformed and reconfigured according to the historical exigencies of the context within which it is operating. As such, the category of race can be seen to be critically reoriented in the context of the early nineteenth century, at a time, uncoincidentally, when European imperialism was extending its reach across the globe. In his analysis of the early nineteenth century, Foucault (2003, 61) identifies a decisive break with the past in relation to the uses and abuses of race and the "discourse of race struggle":

> It [the discourse of race struggle] will become the discourse of a centered, centralized, and centralizing power. It will become the discourse of battle that has to be waged not between races, but by a race that is portrayed as the one true race, the race that holds power and is entitled to define the norm, and against those who deviate from that norm, against those who pose a threat to the biological heritage. At this point, we have all those biological-racist discourses of degeneracy, but also all those institutions within the social body which make the discourse of race struggle function as a principle of exclusion and segregation and, ultimately, as a way of normalizing society.

Foucault (2003, 60) identifies the resultant "race wars" that this discourse of race struggle enables as what is "articulated with European policies of colonization." And it is precisely at this biopolitical juncture that biometrics, as understood in the contemporary sense, emerges.

Pearson's Biometrics, Race Struggle and the Biopolitics of Biologically Variable Subjects

In his tracking of the emergence of biopolitics in the late eighteenth century, Foucault (1990, 139) brings into focus two poles that establish its conditions of possibility: first, techniques of power exercised through the *disciplines* and principally concerned with the body, forming "an *anatomo-politics of the human body*"; and, second, what he terms "*a biopolitics of the population*," administered through "an entire series of interventions and *regulatory controls*"; biopolitics "focused on the species body, the body imbued with the mechanics of life and serving as the basis of the biological processes: propagation, births and mortality, the level of health, life expectancy and longevity, with all the conditions that can cause these to vary." Encoded in the very nomenclature *biometrics*, in its nineteenth-century form and prior to its contemporary resignification, is precisely this biopolitical range of meanings. In its nineteenth-century context, biometrics signified "the application of modern statistical methods to the measurement of biological (variable) objects" and the application of "statistics to the problem of biology" (*Compact Oxford English Dictionary* 1992). Under the aegis of Karl Pearson (1857–1936), founder of the Biometric Laboratory, University College, London, and of the journal *Biometrika*, biometrics was driven by an unequivocal biopolitical agenda. Pearson's development of biometrics was underpinned by the intertwining of race, eugenics, mathematics and theories of social Darwinism. Biometrics, for Pearson, became the application of mathematics to the analysis of life forms and what for him were pressing questions of population "stock":

> In Pearson's view, the imperial nation required more than an economic framework designed to give its citizens a material stake in its power; it also demanded the "high pitch of internal efficiency" won by "insuring that its numbers are substantially recruited from the better stocks"
>
> (Kevles 2004, 32)

Unfortunately, Daniel Kevles truncates this excerpt from Pearson's *National Life from the Standpoint of Science* (1901). I cite the excerpt in full:

> My view—and I think it may be called the scientific view of a nation—is that of an organized whole, kept up to a high pitch of internal efficiency by insuring that its numbers are substantially recruited from the better stocks, and kept up to a high pitch of efficiency by contest, chiefly by way of war with inferior races, and with equal races by the struggle for trade-routes and for the sources of raw material and of food supply. This is the natural history view of mankind, and I do not think you can in its main features subvert it.
>
> (Pearson 1901, 43–44)

In Pearson's "natural history view" of the world, race, capitalism, empire and war all function to constitute his scientific vision of a society working at its optimum pitch. Pearson's biometric view of "better stocks" was fundamentally informed by whiteness, its racial hierarchies, and the "race struggle" that Foucault identified as crucial to biopolitical programs:

> What I have said about bad stock seems to me to hold for the lower races of man. How many centuries, how many thousands of years, have the Kaffir and the Negro held large districts in Africa undisturbed by the white man? Yet their inter-tribal struggles have not yet produced a civilization in the least comparable with the Aryan. Educate and nurture them as you will, I do not believe that you will succeed in modifying the stock. History shows me one way, and one way only, in which a high state of civilization has been produced, namely, the struggle of race with race, and the survival of the physically and mentally fitter race. If you want to know whether the lower races of man can evolve to a higher type, I fear the only course is to leave them to fight it out amongst themselves, and even then the struggle for existence between individual and individual, between tribe and tribe, may not be supported by that physical selection due to a particular climate on which probably so much of the Aryan's success depended.
>
> (Pearson 1901, 19–20)

Inscribed in Pearson's formalisation of biometrics into a scientific discipline, concerned with the application of mathematics and statistical methods to the measurement of biological (variable) objects and racial "stocks," is a biopolitics of race driven by the need to protect the white nation from the degenerations produced by inter-racial crossings: "Frequently they [races] intercross, and if the bad stock be raised the good is lowered" (Pearson 1901, 20). Informing Pearson's concept of race is the notion of "racial purity, with all its monistic, Statist, and biological implications" (Foucault 2003, 81). Enunciated in Pearson's fear of racial intermixing is an exhortation to the state to be "the protector of the integrity, superiority, and the purity of the race" (Foucault 2003, 81). In his scientific address to the British nation, Pearson (1901, 21) offers a solution that is unapologetically colonial and necropolitical: "The only healthy solution is that he [the white man] should go [to the lands of the inferior races], and completely drive out the inferior race." Pearson is here operating in what Foucault (2003, 257) terms the "biopower mode," an exterminatory mode invested in mobilising race in order to legitimate what he terms "colonizing genocide." In the context of the research conducted by Pearson and his colleagues at the Biometric Laboratory, University College, London, this exterminatory mode of biopower was conceptualised as something that needed to be exercised not just on "inferior races" but on an entire spectrum of human variation that included people marked as "deviations" from heteronormative sexualities and those classified as "unfit" because of their "abnormal" physical or mental attributes. As Lennard Davis (2006, 9) writes, Pearson's Biometric Laboratory "gathered eugenic information on the inheritance of physical and mental traits including 'scientific, commercial, and legal ability, but also hermaphroditism, hemophelia, cleft palate, harelip, tuberculosis, diabetes, deaf-mutism, polydactyly (more than five fingers) or brachydactyly (stub fingers), insanity, and mental deficiency' ... All these deviations from the norm were regarded in the long run as contributing to the disease of the nation"; these "deviations" were precisely what needed to be eugenically "bred out" of the corpus of the white nation.

In the traditional histories of biometric technologies (for example, Jain and Ross 2008; NSTC 2006), Pearson does not figure, despite the fact that he is credited with formalising a scientific discipline, biometrics, that later lent its nomenclature to the technologies that followed. I have attempted to re-insert him within my genealogy of biometrics not only because he established a scientific discipline under that rubric, but because his investment in biometrics was fundamentally shaped by biopolitical agendas that have been and continue to be constitutive of biometrics.

Colonial and Metropolitan Biopolitical Apparatuses: Biometrics and Regimes of Veridiction

In a brief reflection on the writing of genealogical histories, Foucault (2003, 10) remarks that what often is produced are "disordered and tattered genealogies." True to spirit, my genealogy of biometrics is "tattered" in its attempt to bring together fragments that would otherwise be excluded (Pearson's biometrics) and "disorderly" in its shuttling, historically, backward and forward along the complex network of relations that constitutes its conditions of emergence. In keeping with the non-linear and non-teleological tracing of history that the genealogical method enables (in order to establish points of connection between seemingly disparate topics and subjects), the formalisation of the discipline of biometrics in the late nineteenth century actually post-dates the development of what would be, in contemporary terms, identified as biometrics "proper": the inked fingerprint as a mark of identification. The inked fingerprint as biometric identificatory marker pre-dates the scientific discipline of biometrics (as the application of mathematics to biology) even as its conditions of emergence are constituted by the same biopolitical concerns of empire, colonial governance, race war and its "principle of exclusion and segregation" (Foucault 2003, 61). What I am marking here is the complex terrain of heterogeneous histories, technologies and disciplines that is inscribed with relations of recursivity and folds. The biometric of the inked fingerprint was not named and officially identified by the term "biometric." After the fact of its emergence, implementation and dissemination, this technology is resignified and absorbed into the fully established fold of biometrics. Operative here is a genealogical dynamic of history marked by the torsions of recursive recuperations and reinscriptions that connect past to present and vice versa. To a degree, in this history, fingerprint biometrics is retrospectively positioned as a sort of ur-biometric and has been invested with an iconic origin-status becoming, in the process, "the de facto international standard for positively identifying individuals" (U.S. General Accounting Office 2002).

In his history of fingerprinting, Simon Cole (2002, 63) observes that:

> Although most people associate fingerprinting with that bastion of modern policing, Scotland Yard, the British system of fingerprint identification actually emerged in the colonies rather than in England, in response to the problem of administering a vast empire with a small corps of civil servants outnumbered by hostile natives.

Cole proceeds to unfold the complex matrix that established the conditions of emergence for forensic fingerprint identification. It is a matrix intersected by issues of colonial rule, law, biological racism and the subjugation of native, anti-colonial insurgents. This matrix, furthermore, is magnetised by the exercise of British violence at a crucial juncture in the imperial occupation of India:

> The birth of modern fingerprint identification came, not coincidentally, at one of the tensest moments in the history of British India. In 1857 Indian conscripts, known as "sepoys," spurred by rumors that the grease that lubricated their rifle cartridges contained beef and pork fat—thus violating the dietary laws of both Hindu and Muslims—rebelled against their British officers and, for a time, took control of Delhi.
>
> (2002, 63–64)

The Sepoy massacre that followed soon re-established the colonial rule of law. But, as Cole (2002, 64) notes, "The Mutiny heightened 'the need to enforce law and order in the unruly colonies more severely,' and, at the same time, it exposed the need for Britain unequivocally to demonstrate its imperial mission in the context of its "barbarous" and unruly colonies.

Faced with this colonial burden and the Orientalist problematic of British officials not being capable of differentiating "one Indian from another," William Herschel, chief administrator of a district in Bengal, proceeded to develop the biometric of inked fingerprint identification in order to establish a reliable system that would at once differentiate and identify his Indian subjects (Cole 2002, 65).

As a technology administered by "the one true race" in order to win the war in the colonial race struggle, fingerprint biometrics was underpinned by biological theories of race and racial hierarchies that constructed categories of innately "'criminal tribes' ... predisposed to criminal behavior"

(Cole 2002, 67). Identified as biologically predisposed to criminality, they could now be categorised as biocriminals and thus they could be summarily isolated, punished and executed. Operative in the colonial classification and biometric identification of "criminal tribes" is the legitimating spectre of what Foucault (2003, 314) calls the "background body" or the "ancestor's body" that functions as a type of "metabody": the inherent criminality of these tribes is causally overdetermined by this hereditary metabody. The role of fingerprint biometrics, in this colonial scene of biopower, is to individuate in the form of somatic singularities the constituent, living parts of this metabody of criminality so that they can be surveilled, regulated and disciplined.

Fundamentally driving the development of the biometric technology of fingerprinting was a shift in the type of question posed by the state to its target subjects. As discussed in the Introduction, the emergence of the biopolitical state, as Foucault (2008, 34) underscores, is marked by the installing of the veridictional question, "the question of truth," at the core of its criminological operations. The question: "What have you done?" is now replaced with the question: "Who are you?" (Foucault 2008, 34). This is the biopolitical question *par excellence*, as it presupposes that who you are will fundamentally determine not only what you have done, but also what you are capable of doing. Thus, in the British colonial context, to be biometrically identified as belonging to an identifiable Indian "criminal tribe" is already to categorise you as "predisposed to criminal behaviour" both in a retrospective tense (from your birth, you have been innately criminal) and in the tense of the future anterior (in the future, you will have already committed a criminal act because of your biological predisposition). This foundational biopolitical question perfectly dovetails with the animating question of all biometric technologies. Biometric systems are premised precisely on answering the veridictional question—"Who are you?"—through a process of either verifying or authenticating a subject's identity by comparing a subject's sample biometric against a database of relevant biometric templates. "With identification," write Woodward et al. (2003, 7), "the biometric system asks and attempts to answer the question, 'Who is X?'" The manner in which the veridictional question of biopolitics is embedded within a biometric conceptualisation of identity as innate (ontologised) is succinctly evidenced by this formulation of biometrics: "What you are: Biometrics" (Woodward et al. 2003, 7). What I am attempting to emphasise here is how the development of biometric technologies (their animating logic, procedures of technological rationality, techniques of subject constitution and so on) was indissociably tied to the emergence of biopolitical forms of governance.

The emergence of fingerprint biometrics, within this field inscribed by race, racism, biology, regimes of veridiction, colonialism and the power to kill, evidences biometric technology's intimate relation with the modern biopolitical state. "It is at this moment," Foucault (2003, 254) writes, "that racism is inscribed as the basic mechanism of power, as it is exercised in modern States. As a result, the modern State can scarcely function without becoming involved with racism at some point, within certain limits and subject to certain conditions." Situated in the stratified genealogy of technologies of body measurement and identification, biometrics emerges as one technology put to the service of the colonial state in order to consolidate and extend its rule over subject peoples. The mathematicisation of the *bios*, upon which biometrics is premised, invests the technology with the cachet of scientificity; this scientificity, underpinned by the principles of mathematical objectivity and technological reproducibility, invests biometrics with a sense of authority and credibility that is invaluable to the everyday operations of the colonial state in the governance of its (hierarchised and segregated) populations.

In the operations of the biopolitical state, racism and its instrumentalising technologies serve to map and consolidate a series of hierarchised relations that enable the segregation of target populations and, in the form of "necropower" (Mbembe 2003), authorise the right to kill with impunity. The hierarchy of races, administered through multiple biopolitical technologies such as fingerprint biometrics, "is a way of separating out the groups that exist within a population. It is, in short, a way of establishing a biological-type caesura within a population that appears to be a biological domain ... That is the first function of racism: to fragment, to create caesuras within the biological continuum addressed by biopower" (Foucault 2003, 255). In the field of nineteenth-century biometrics and anthropometrics, the biological-type caesura is enunciated by the "inferior whorls" of a fingerprint, by the size of a cranium, and so on. From these entirely local and corporeally situated loci flow a series of macro-political effects of biopolitical governance and control. The use of biometric technologies in order to surveil, identify and control colonised subjects flourished during the twentieth century, under Benito Mussolini's fascist regime, in the context of Italy's colonies in North Africa (Gibson, 2002 148), with the implementation of biometric dossiers on Italy's colonised subjects. And, as I discuss in Chapter 3, the use of biometric technologies in the service of colonial occupation continues apace with the U.S. military's deployment of biometrics in the context of the war in Iraq.

The biopolitical problem of governing and keeping under control suspect populations was one that did not occupy European powers solely in the context of their respective colonies. It was a problem that also preoccupied European governments within the dispersed spaces of their metropolises. The classification of suspect populations—including such categories as "natives," "the labouring class," "deviants," "the mad," "prostitutes," and so on—was due to the emergence of a cluster of social sciences (that included Lombroso's criminal anthropology, Galton's eugenics, and Pearson's biometrics) preoccupied with biopolitical problems (such as race mixing and degeneration, criminality and hereditary traits). These new social science disciplines, as biopolitical apparatuses, worked in tandem with key government institutions, including the police, thereby "guaranteeing relations of domination and effects of hegemony" (Foucault 1990, 141). It is at this historical juncture that the biometric of the inked fingerprint is exported from the colonies to metropolitan Europe, where it is incorporated as a key biopolitical technology in the identification, surveillance and disciplinary control of Europe's internal others, producing, in effect, a type of "internal colonialism." "It should never be forgotten," Foucault (2003, 103) writes,

> that while colonization, with its techniques and its political and juridical weapons, obviously transported European models to other continents, it also had a considerable boomerang effect on the mechanisms of power in the West, and on the apparatuses, institutions, and techniques of power. A whole series of colonial models was brought back to the West, and the result was that the West could practice something resembling colonization, or an internal colonialism, on itself.

First formulated in the far-flung outpost of the British empire in order to solve a specific problem of biopolitical-colonial governance (the inability of British officials to distinguish between their Indian subjects), the biometric of the inked fingerprint proved to be an invaluable technology in the biopolitical governance of Europe's internal others, in an historical context marked by rapid economic and historical change. The individuated body constituted by the disciplinary apparatus of the inked fingerprint is, in this context, enmeshed within larger configurations of biopower concerned with the surveillance and regulation of target populations. Foucault (2003, 240–50) articulates that charged point of intersection at which disciplinary power becomes imbricated with biopower:

> It is as though power, which used to have sovereignty as its modality or organizing schema, found itself unable to govern the economic and political body of a society that was undergoing both a demographic explosion and industrialization. So much so that far too many things were escaping the old mechanism of the power of sovereignty, both at the top and at the bottom, both at the level of detail and at the mass level. A first adjustment was made to take care of the details. Discipline had meant adjusting power mechanisms to the individual body by using surveillance and training ... And then at the end of the eighteenth century, you have a second adjustment; the mechanisms are adjusted to phenomena of population, to the biological or biosocial processes characteristic of human masses.

Operating in tandem, then, are two modalities of power: "the body-organism-discipline-institutions series, and the population-biological-processes-regulatory mechanisms-State" (Foucault 2003, 250). The French police clerk, Alphonse Bertillon (1853–1914), was instrumental in the development of a biometric technology, police anthropometry, that was at once focused on the exercise of disciplinary power on the body-organism of the individual in the institutional context of the police station *and* on the larger, biosocial problem of the state in identifying and regulating criminal populations of "recidivists." Police anthropometry, as Martine Kaluszynski (2001, 123) argues, "was not simply a new weapon in the armory of repression, but a revolutionary technique: it placed identity and identification at the heart of the government policy, introducing a spirit and set of principles that still exist today." In 1883, Bertillon announced that he had developed a filing system that, through the use of anthropometric measurements and forensic photography, classified and identified criminals and, in particular, recidivists. Naming his biometric system "anthropometrical *signalment*," Bertillon subjected the criminal body to a series of eleven measurements that focused on those parts of the body that could "most easily be measured accurately," including "the length and breadth of the head and of the right ear, the length of the elbow to the end of the middle finger, that of the middle and ring fingers themselves; the length of the left foot, the height, the length of the trunk (*buste*), and that of the outstretched arms from middle to middle finger-end" (Lalvani 1996, 109).

Aside from identifying and classifying a standardised series of bodily measurements, Bertillon, in an innovative move, developed photography's potential as a forensic technology of the body by implementing a series of standardised procedures that would regulate and discipline how the criminal body was photographed: "the focal length was standardized and the body of the criminal exposed to an 'even and consistent lighting.' Furthermore, the problem posed by facial expression, which had thwarted previous attempts at photographing criminals, was neutralized by the use of the profile view" (Lalvani 1996, 109–13). Bertillon's system of "anthropometrical *signalment*" can be seen to be one of the most important forerunners of contemporary biometric systems precisely because he instituted a system whereby he transmuted individual body measurements into data that was, in turn, stored on cards in extensive filing systems that classified the captured anthropometric attributes of criminals along a complex series of subdivisions. It was the very complexity of Bertillon's system that caused it, in the end, to fail. The inked fingerprint of classification and identification, developed in 1897 by Azizul Haque, assistant to Edward Henry, chief of police in Bengal, soon became the biometric system of choice for law enforcement agencies in Britain (1901), the U.S.A. (1904) and around the world (Beavan 2001; Cole 2002). Bertillon had, however, inaugurated a system of bodily measurement and classification that would, in the next century, be incorporated in the development of digital databases devoted to the classification and identification of criminal bodies. Suren Lalvani (1996, 116) draws attention to the ramifications of Bertillon's anthropometric filing system:

> the body becomes permeable to the principle of surveillance and the gaze of power flickers over these bodies, isolating, neutralizing, standardizing and reducing them, one and all, to the topography of power. Power makes of the signs of the criminal body a textual practice. The body is objectified, divided, analyzed and organized into a cellular structure of space (the file index) – the representative architecture of surveillance; it is individuated, transformed into a subject and subjected, a discursive object within a disciplinary apparatus.

The file index emerges as foundational to the operations of biopolitical apparatuses, at once capturing the body, rendering it "docile and forced to yield up its truth" (Tagg 1988, 76), and transposing it to networked relations of power. The nineteenth-century file index emerges as the instrumental template for the development and expansion of biopower. The file index, in its classifying and identifying of a subject's corporeal attributes, is a technology of individuation that, at a micro level, produces the subject as a somatic singularity; at a macro biopolitical level, the file index functions as a modality of dispersion, enabling the dissemination of the individuated subject through networked and interoperable databases.

Bertillon's system of storing and classifying biometric data within a filing system was, by the 1930s, greatly streamlined by the FBI's use of an IBM punch-card sorter "that could retrieve all fingerprint cards containing a certain classification" (Cole 2002, 251). In his tracking of the increasing use by police authorities of computers, with their extraordinary power and speed to store, classify and deliver biometric data, Cole (2002, 252) observes that: "By 1983, The FBI reported, 'the total fingerprint file of all criminals born after 1928 was on-line, and all searches were routinely done in this new system'." Contemporary biometric technologies emerge at the juncture of biopolitical concerns, the development of the surveillance society (Lyon 2004, 2008, 2008a) and the take-off of computers in the 1960s-70s (Parenti 2003). Authorities such as the FBI and the U.S. Department of Defense have played instrumental roles in funding the development of biometrics (NSTC 2006). It is at this historical juncture that continuities and ruptures mark this genealogical tracking of biometrics that I have attempted.

On the one hand, the nineteenth-century biopolitical preoccupation with identifying, classifying and governing "suspect" populations continues (for example, in the tracking of "terrorists," "insurgents" and so on). On the other, the digital revolution can be seen to instantiate a break with the past. In its capacity to transmute corporeal matter into digitised data, corporeal attributes now become coextensive with *in silico* matter. Operative here is a "conception of information as (disembodied) entity that can flow between carbon-based organic components and silicon-based electronic components to make protein and silicon operate as a single system" (Hayles 1999, 2). The formative specificity of the medium in which information is materialised is here both disavowed and effaced. A type of "informatic essentialism" ensures that the body "is therefore subject to the same set of technical actions and regulations as in all information" (Thacker 2003, 86). What I am marking here is the manner in which the body is now subject to an intensification of instrumentalising techniques and procedures. As digitised bits of information, the body-as-information can now be inserted within networked relations of biopower that traverse the local, the national and the global. The purchase on identity, in this digital landscape, has lost none of its biopolitical salience or power. On the contrary, as I discuss in Chapter 3, it has morphed into the military-imperial quest for "identity dominance."

Biometric systems, once situated within this context, function as exemplary technologies of biopower. Emerging from a long and complex history of submitting the body to mathematical measurements in order to determine identificatory attributes always charged with political investments, biometric systems are predicated on the notion that the body can be subject to disciplinary economies of explicit calculations, classification, surveillance and control. The body, when screened by biometric technologies, is divided into corporeal components (the iris, the fingerprint, the face) that are individually processed and converted into algorithmic formulae and then stored as templates within biometric databases. Individual body parts are, in this process, inscribed within anatomies of biopower that enable the operations of regimes of identification, classification and disciplinary normativity. Biometric technologies at once are capable of capturing the entirety of the body (through gait signature biometrics), its surfaces (facial and finger scans), its depths (vein recognition biometrics) and its emanations (odour-sensing biometrics), "making it possible," in Foucault's (1982, 216) words, "to bring the effects of power to the most minute and distant elements. It assures an infinitesimal distribution of the power relations." As I discuss in Chapter 3, this infinitesimal distribution of power relations has been effectively facilitated by the establishment of networked, multi-modal, interoperative biometric systems.

2 The Biometrics Of Infrastructural Whiteness

Technologies of Capture, White Templates and Coloured Occlusions

In tracing a genealogy of biometrics, I argued in the previous chapter that race has been fundamentally constitutive of the operations of biometric technologies. This chapter is concerned with examining the point of intersection between biometric technologies, bodies and race in the context of a phenomenon termed biometric "failure to enrol." Contemporary biometric systems employ technologies that scan a subject's physiological or behavioural characteristics in order to verify or authenticate their identity. Biometrics can thus be succinctly characterised as a technology of capture: that is, the technology is fundamentally predicated on capturing images of subjects. It is this process of visual capture that enables the processes of template creation and consequent verification or authentication of a subject's identity. Situated in the context of this paramount concern with the visual capture of a subject's identificatory characteristics, this chapter is concerned with those very instances where biometric systems fail to capture a subject's image—precisely because of their race, precisely because the subject fails to conform to predetermined white standards that set the operating limits of particular biometric technologies.

Animating these racialised failures of representation is the power of whiteness. Whiteness is a racial category like no other. Its power appears to spring from a phantasmatic quality that is simultaneously grounded in the very concrete and mundane infrastructure of everyday life (see, for example, Fanon 1970[1952]; Frankenberg 1993; Dyer 1997; and Thandeka 2000). At this level, whiteness is so infrastructurally diffuse as to be imperceptible—that is, whiteness, as I will proceed to demonstrate, so constitutes the molecular fabric of everyday technologies and practices that it cannot appear as a racial category as such. It is precisely because whiteness is so enucleated into the material weave of everyday life that one cannot talk of whiteness *as such*. No *as such,* so to speak, because the power of whiteness resides in this capacity to occlude and so mystify its status as a racial category that it too often escapes taxonomic determination, while simultaneously remaining the superordinate racial category that effectively determines the distribution of all other classificatory categories along the racial scale.

In the course of this chapter, I will ground this seemingly abstract argument on the power of whiteness in the context of contemporary biometric technologies of identification and verification. It is within this specific context that I discuss how non-white subjects are, according to the technical literature, often precluded from biometric enrolment due to the fact the technologies fail to "read" their biometric characteristics. I will argue, in other words, that a number of these biometric technologies are *infrastructurally calibrated to whiteness*—that is, whiteness is configured as the universal gauge that determines the technical settings and parameters for the visual imaging and capture of a subject. I will examine the operation of the racial category of whiteness in the context of three particular biometric technologies: facial-scan, finger-scan and iris-scan systems.

Infrastructural Whiteness

In "The light of the world," a chapter concerned with disclosing the manner in which the racial category of whiteness informs the technologies of photography and film, Richard Dyer (1997, 83) argues that "All technologies are at once technical in the most limited sense (to do with their material properties and functioning) and also always social (economic, cultural, ideological)." Dyer (1997, 89) then proceeds to track the manner in which "photographic media and, *a fortiori*, movie lighting assume, privilege and construct whiteness." Focusing on the complex interplay of various technological elements, including film stock, different types of lighting, and camera apertures, Dyer (1997, 90) demonstrates how technologies of photography and film "were developed taking the white face as the touchstone." In the process, Dyer (1997, 89) explains why, for instance, in school photos "the black pupils' faces look like blobs or the white pupils have theirs bleached out."

I want to transpose Dyer's illuminating analysis of the racialised, specifically white, elements that inform the technologies of photography and film to the imaging technologies of digital facial scans and finger scans. Before I proceed down this track, however, I want to take a moment to problematise Dyer's conceptualisation of whiteness. Throughout his book *White*, Dyer deploys a concept of whiteness that is predicated on a totalising, ahistorical and essentialised understanding of the category. This is succinctly encapsulated in Dyer's (1997, 4) argument that what he "is studying [is] whiteness *qua* whiteness ... whiteness itself." As I have argued elsewhere, in deploying such a totalising and ahistorical conceptualisation of whiteness, Dyer proceeds to range freely across a wide spectrum of historical contexts, genres and media and, in the process, generates an anachronistic schema in which, for example, a fifteenth-century painting by Bellini participates in the same symbolic articulation of whiteness as does Sylvester

Stallone in *Rambo* (Pugliese 2002). In the discursively untenable move of situating, under the ahistorical rubric of whiteness, a Bellini painting in a politically and ideologically equivalent relation to *Rambo*, Dyer effectively erases, for example, that complex genealogy that marks the fraught relationship of Italian immigrants to the category of whiteness in the U.S. and in Australia (see Gambino 1974; Roediger 1994; Jacobson 1998; Pugliese 2002; Guglielmo and Salerno 2003).

In examining whiteness in contemporary biometric technologies, I want to pose whiteness in *infrastructural* terms, that is, as an element which is indissociable from the effective operations of a particular technology. In posing whiteness as infrastructural, I am not suggesting that this racial category is some sort of ahistorical and essentialised datum, what Dyer terms "whiteness *qua* whiteness"; rather, I will be arguing that whiteness must be read in terms of a racial category that is historically situated, marked by the specificity of particular media and technological apparatuses, and calibrated by identifiable discourses, laws and conventions. In talking of an infrastructural whiteness, I will be drawing attention to the very *structurality* of its infrastructure; in other words, I want to bring into focus what effectively gets invisibilised when technologies are represented as ideologically neutral "conduits" of data, rather than ideologically inflected constructors of knowledge/power.

If whiteness is to be invested with any power, it must be capable of a potentially infinite process of situated, historical repeatability. Couched in Derridean terms, Dyer's conceptualisation of "whiteness *qua* whiteness" is "in itself divided and multiplied in advance by its structure of repeatability" (Derrida 1990, 48). Viewed in strictly rhetorical terms, the figure of *diacope* (repetition of a word [whiteness] with one word [*qua*] in between) constitutes its logic of signification, as it already underscores its (infra)structure of repeatability and its openness to alterity with every instance of transposition/iteration across diverse media and contexts. In other words, the very *qua* of whiteness, its assumed essence, is dependent upon "its structure of repeatability," where its every iteration entails that "something new takes place" (Derrida 1990, 40).

I stage this brief deconstruction of Dyer's essentialised concept of whiteness not in order to indulge in a series of rhetorical flourishes but to underscore the manner in which the power of whiteness resides in the fact that it is never, because of its very structure of repeatability, essentially identical to itself. In not being strictly identical to itself, while simultaneously being capable of potentially infinite iterations, whiteness can be seen to be invested with a power historically to mutate, adapt and, in the process, arrogate different technologies, bodies, races and ethnicities in its situated repetitions. If this colonising flexibility and imperial inventiveness constitutes the power of whiteness as a racial category, then it also exposes whiteness to risk. Precisely in not being identical to itself because of its (infra)structural iterability, whiteness risks dissolving those very pliable borders that enable its "flexible positional superiority," to draw on Said's (1991, 7) apposite phrase. This marks, in other words, the urgent need always to put in place legislation (for example, the White Australia Policy [Jupp 1991]), laws (for example, the "one drop" of black blood rule in the U.S. [Davis 1991; López 1996]) and other regulatory mechanisms designed to control and govern its categorical purity in the face of historical forces and agents that may attempt to contest, contaminate and miscegenate its illusory pristine status.

Subjects of Irreflectivity: Biometrics' Occlusion of Coloured Bodies

In order to enrol within a biometric system, the subject whose identity will be verified by the particular biometric system is required initially to supply the requisite biometric data, such as a digital scan of their face, which is subsequently converted into a template. The template, which is generated by the algorithmic encoding of a subject's distinctive biometric features, is stored in the system and is used to verify a user's identity every time they present themselves for biometric screening—in other words, the initial enrolment template is matched against the user's verification template.

It would seem that the process of biometric enrolment is a straightforward process: subjects present themselves to a biometric system; their biometric data are extracted and algorithmically converted into a template that is consequently used for either verification or identification. Yet, within the biometric industry, there is also what is termed "failure to enrol [FTE]" whereby certain subjects' features cannot be "extracted" or "acquired" by the relevant biometric systems. FTE is due to what is known as "Failure to Acquire (FTA) also known as Failure to Capture (FTC)," and it "denotes the proportion of times the biometric device fails to capture a sample when the biometric characteristic is presented to it" (Jain and Ross 2008, 10). Significantly, this failure to enrol is neither random nor arbitrary. Rather, it is marked by the fact that only certain ethnic or demographic groups appear to experience this phenomenon. "Certain ethnic and demographic populations," write Nanavati et al. (2002, 35–36), "are more prone to high FTE rates than others ... Those of Pacific Rim/Asian descent are more prone to FTE than control groups ... Users of Pacific Rim/Asian descent may have faint fingerprint ridges—especially female users."

This failure to enrol occurs across a number of biometric systems, including finger-scan, iris-scan and facial-scan technologies. I want to focus specifically on FTE in the context of finger-scan and facial-scan technologies. "Testing of facial-scan solutions indicates," write Nanavati et al. (2002, 37), "that the technology may not be as adept at enrolling very dark-skinned users. The increased FTE rate is not attributable to the lack of distinctive features, of course, but to the quality of the images provided to the facial-scan systems by video cameras optimized for lighter-skinned users." Despite the acknowledgement that FTE does not result because dark-skinned users "lack distinctive features," the fact that biometric technologies might be "optimized for lighter-skinned users" still fails to prompt the authors to proceed to name the constitutive role of whiteness as an infrastructural racialised gauge that sets the operating parameters of these image acquisition technologies.

Nanavati et al. (2002, 65–66) are attentive to the question of lighting/race without ever unpacking the larger ramifications of this powerful nexus: "facial-scan technologies are generally unable to acquire images that are somewhat overexposed or underexposed"—we are here in the realm of the "black blobs" and "white bleachings" that Dyer (1997, 89) draws attention to in his mapping of the effects of this nexus. From this point on, Nanavati et al. (2002, 66) continue to sidestep the problematic of whiteness as an infrastructural racialised gauge:

> Facial-scan systems' sensitivity to lighting and gain can actually result in reduced ability to acquire faces from individuals of certain races and ethnicities. Select Hispanic, black and Asian individuals can be more difficult to enrol and verify in some facial-scan systems because acquisition devices are not always optimised to acquire darker faces. At times, an individual may stand in front of a facial-scan system and simply not be found. While the issue of failure-to-enrol is present in all biometric systems, many are surprised that facial-scan systems occasionally encounter faces they cannot enrol.

The articulation that "many are surprised" at this FTE regarding facial-scan systems marks a double moment of occlusion: the systemic, empirical occlusion of the non-white face before the biometric system calibrated to the white gauge, and the ideological occlusion that it is this very white calibration of biometric systems that precludes the acquisition of the features of non-white subjects. This moment of occlusion must be named in terms of a racialised blind-spot, a technological and discursive point of irreflectivity that cuts in two directions at once: failure to begin to theorise on the technological/race nexus of these systems (thus the "surprise" that some subjects are "simply not found") and the failure of non-white bodies to function as reflective subjects that emit sufficient light to register precisely as template subjects of enrolment in the face of biometric systems whose image acquisition parameters are predetermined by an infrastructural whiteness. Everything from this point gestures towards a racialised zero degree of non-representation. And this racialised question of a subject's epidermal reflectivity is not something merely confined to the field of biometrics. On the contrary, it occupies a much larger social domain and it is sometimes generative of fatal effects. Mmaskepe Sejoe (2009) has discussed how black people in Australia's Northern Territory, with a large Aboriginal population, are often sarcastically referred to by whites as "non-reflectives"; the point to this racist slur being that if a black person is run over at night, then they are at fault for being "non-reflectives." Gracelyn Smallwood (2009) has documented how, one evening, an Aboriginal woman was hit by a car driven by a white driver in a car park and left to die, despite the fact he was aware that he had run her over; the woman took four hours to die alone in the car park. The white driver's excuse was that she was a "non-reflective" and that he thus had not seen her; he was let off with a $750 fine.

In his discussion of the schematised body in the history of visual culture, James Elkins (1999, 279) argues that "schemata are denials of the body," and as a result "whatever is taken to be properly not an attribute of the visualized body ... is excluded from representation, finessed, glossed, or otherwise inadequately or partially shown." Transposing Elkins' argument to the schematised white bodies that operate as image-acquisition templates in certain biometric technologies, non-white corporeal features are, in anthropometric terms, disproportionate to the norms of the white body. Viewed in the context of da Vinci's founding metaphor of the white body as symmetrical analogue to the perfection of Euclidean geometry (as discussed in Chapter 1), where the square and the circle constitute the geometric matrix from whence emerges a perfectly symmetrical body—the non-white body's corporeal differences signify as so many excrescences that fail to conform to the smooth and homogenised body of white schemata. As such, "non-normative" racial differences are what must be finessed away in order to configure the template, universal white body. Black or coloured skin, "Asian" fingerprints and so on are some of the physical attributes that are marked by a visual excess that transgresses the law of the body proper. These racialised body-bits figuratively fail to square the circle of the template body. That is why they must be finessed away from the white template body, only to be "discovered" out there in the disordered world of racialised corporealities that exceeds the normativity of white schematic borders and limits. The sudden discovery of racial difference in the texts of biometrics through its failure to be technologically registered ("Look, black skin and the phenomenon of failure to enrol") is what generates the sense of the unexpected naming of race and ethnicity in the seemingly neutral texts of science and technology. Encoded in these abrupt namings of (non-white) race is that haunting Fanonian echo: "Look, a Negro!"—and the ensuing "objective examination" and "discovery of blackness ... All this whiteness that burns me" (Fanon 1970, 79, 81). Fanon here counterposes the indelible facticity of blackness, sealed in its "epidermal schema," to an incendiary whiteness that disavows its own racial status even as it sears into being the raced reality of the other.

In this schema, non-white corporeal differences are marked doubly by an exteriority: they are "out there" in the messy world of carnal bodies and, simultaneously, they are also exterior to the self-contained internal symmetry of the white template body. By virtue of their excessive and exorbitant features, non-white bodies are, by definition, hyperbolic bodies: they are always already *troppo*. In the context of the geometrically circumscribed world of the template white body, these other bodies are tropically off the map. In the context of Eurocentric representations of the body, the history of these "other" bodies is perhaps emblematically illustrated by Saartjie Baartman, the so-called Hottentot Venus, exemplar of a body that violated the canon of white corporeality and that, as "biological curiosity," confirmed for whites the fact that "blacks were aliens" (Arogundade 2000, 14).

The visual semiotics of the caucacentric template body, as schema, are predicated on the excision and exclusion of non-white phenotypical features. Yet, even as these non-white phenotypical features are excluded, they must stage their return by being grafted onto the white schematised body when the biometrician is brought face-to-face with the non-white body. For these non-white adjuncts (black skin, "Asian" fingerprints) to be grafted onto the white template, they must be imported back from the outer-limits of representation where they have been relegated. This other place is best described, drawing on an apposite term from film theory, as the "space-off": "the space not visible in the frame but inferable from what the frame makes visible" (de Lauretis 1987, 26). The space-off is where the non-white body is located: excluded and foreclosed from the schematic contours of the white figure, it is inferable from what the template white body makes visible: *racial difference*.

"Human-Free" Technology: Invisibilising the Template White Body in the Biometric Machine

The racialised zero degree of non-representation that is instantiated by FTE of people of colour must not be viewed in terms of some mystifying technological "anomaly" or "glitch." On the contrary, I would argue that particular biometric technologies are infrastructurally *calibrated to whiteness*—that is, whiteness is configured as the universal gauge that determines the technical settings and parameters for the visual imaging and capture of a subject. I will draw on the term *calibration* as it effectively encapsulates the three key levels of signification that inscribe the operation of whiteness in biometric technologies. On one level, to calibrate a technology is "to graduate a gauge of any kind with allowance for its irregularities" (*Compact Oxford English Dictionary* [*COED*]). As universal gauge, whiteness is the absolute standard within certain biometric technologies, targeting the capture of white subjects but also allowing for a degree of white variations in skin tone and colour—that is, allowing for certain "irregularities" that may fall outside this standard. These "irregularities," however, are circumscribed within a clearly delineated zone of whiteness and its various chromatic variations. Outside of this diffuse, greyscale zone of whiteness reside non-white subjects who are literally beyond the pale. As I have demonstrated, this degree of allowance literally cuts off when biometric imaging technologies are confronted by subjects whose biometric details—for example, their dark skin colour—are so "irregular" as to fall outside the technical parameters set for image capture. Here the term "irregular" graphically illustrates the disciplinary power of the white gauge in determining the normative standards of imaging technologies. In such instances, the non-white subject, in literally failing to appear before the biometric system despite her or his physical presentation, is dispatched to the outer-limits of non-appearance, to the zero degree of non-representation. *Calibration* perfectly resonates with this imaging economy in that it is a term that effectively belongs to the lexical set of camera settings: a camera operator has a repertoire of calibrations at hand when filming; these calibrations operate at the level of lighting, aperture and lens focus.

On another level, the process of calibrating a technology means not only to establish "a set of graduations or markings," but also to generate a "classification" (COED). The calibration to whiteness of biometric systems, in other words, not only determines the universal gauge of these imaging technologies, it also implicitly reproduces the legendary racial system of classification and hierarchy that places whiteness at the apex followed, in a graduating scale, by Asians and blacks (Banton 1987, 28–31). Within western economies of visual representation, this racial hierarchy has systemically guaranteed the non-representation of non-whites across a wide spectrum of visual media.

Finally, the semantic core of calibration is derived from the term *calibre*. Calibre refers to a "degree of social standing or importance, quality, rank; 'stamp,' degree of merit or importance" (COED). In the context of the calibration to whiteness of biometric technologies, the term calibre underscores relations of power and hierarchy that inflect the physical settings of imaging technologies, as whiteness assumes the gauge of "merit or importance" that determines who may or may not be visually captured within the calibrated zone of representation.

I have spent some time unpacking the infrastructural calibration to whiteness, in particular facial-scan technologies in order to interrogate ongoing doctrinal assertions in the scientific literature that biometric technologies are to be celebrated because of their objectivity and impartiality in processing racial and ethnic subjects. For example, Woodward et al. (2003, 254) argue that:

> The technological impartiality of facial recognition ... offers a significant benefit for society. While humans are adept at recognizing facial features, we also have prejudices and preconceptions. The controversy surrounding racial profiling is a leading example. Facial recognition systems do not focus on a person's skin color, hairstyle, or manner of dress, and they do not rely on racial stereotypes. On the contrary, a typical system uses objectively measurable facial features, such as the distances and angles between geometric points on the face, to recognize a specific individual. With biometrics, human recognition can become relatively more "human-free" therefore free from many human flaws.

The "technological impartiality" of biometric facial recognition can only be maintained by continuing to invisibilise the infrastructural calibration to whiteness that inscribes particular facial-scan systems. Contra Woodward et al. (2003), I would argue that this calibration to whiteness constitutes simply another example of racial "prejudice and preconception" in that it biometrically discriminates between white and non-white subjects. The untenability of arguing that facial recognition systems "do not focus on a person's skin color" is graphically exemplified when one considers that it is precisely a non-white subject's skin colour—specifically, the degree of epidermal and chromatic saturation to blackness—that will determine whether they will be situated outside the operating parameters of a biometric system's image acquisition zone (of whiteness), despite its inbuilt "allowance" for chromatic "irregularities." Woodward et al.'s (2003) invocation of "objectively measurable facial features" and "the distances and angles between geometric points on the face," in order to evidence the impartiality of biometrics, clearly resonates with the language of such discredited nineteenth-century racist scientific disciplines as anthropometry, craniology, phrenology and so on. As discussed in Chapter 1, these racialising and racist disciplines were all predicated on the so-called objectivity of mathematics and geometry in the measurement and classification of human bodies. In the discourses of the sciences, whenever the spectre of race is evoked, inevitably mathematics (or statistics) is mobilised in order to magically transcend the prejudices and preconceptions of the observer that are in danger of contaminating their object of inquiry. Through this sleight-of-hand, the human elements that labour to construct the racialised software of biometric systems are effectively effaced, leaving "human free" geometry to carry out its impartial scanning of subjects.

I want to elaborate on the power of this infrastructural calibration to whiteness in biometric technologies by focusing on finger-scan technologies. In the context of finger-scan biometric systems, marked by the fact "some Asian populations are more likely to be unable to enrol in some finger-scan systems" because they "have lower-quality fingerprints," in particular "faint fingerprint ridges—especially female users" (Nanavati et al. 2002, 60, 37), the bodies of certain Asian subjects are represented as illegible; their "lower quality" "faint" fingerprint ridges are, because of this gendered and racialised infrastructural gauge, finessed beyond schematicity: they literally fail to figure, imagistically and digitally, as templates. That the "lower-quality" or "faintness" of the fingerprint ridges of certain "Asian" or "Pacific Rim" subjects might be the result of a calibration of image acquisition gauges that are set to capture the racial specificities of white subjects must remain, in keeping with the power of white supremacism, unthought. In one stroke, both whiteness (as the constitutive image-acquisition gauge) and the non-white subject (as FTE non-subject) get invisibilised. This process of occlusion and invisibilisation must be displaced to that region of irreflection that construes the problem as intrinsically technological and not also as discursive or ideological in the infrastructural sense.

The social Darwinian resonances of "lower-quality" fingerprints must not be ignored, as they paradigmatically situate Asian bodies on a lower position on that racial hierarchy constituted, respectively, by Caucasian, Mongoloid (Asians), and Negroid races (see Chapter 1). The qualification, "lower-quality," resonates genealogically with the biopolitical use of fingerprints in order to make racialised assessments of target populations. I have in mind here the Norwegian researcher Kristine Bonnevie who, in 1924, after extensive research on the racial dimensions of fingerprints, "found that Asians had relatively more whorls and Europeans had more arches and loops. For a European, this finding accorded well with ... [the] contention that arches were 'more evolved' since it suggested that Europeans were more evolved than Asians" (Cole 2002, 111).

Furthermore, following in the conceptual wake of these massified racial configurations—"Asian" and "Pacific Rim"—what remains to be determined are the specific ethnicities (Thai? Japanese? Samoan?) that constitute the heterogeneity of these totalising categories. Informing the homogenising operations of these totalising categories is a racialising logic that invariably presents itself as self-evident: for example, the essentialised figure of the "Asian" need not be ethnically specified as it is always pre-comprehended and interchangeable: it is enough rhetorically to gesture to the seeming qualification of "certain Asian populations" without having, in practice, to qualify the specificity of the "certain" as such. In the context of this Orientalist logic, driven by the perceived seriality of the figure of "the Asian," such a move would be already redundant. There is, furthermore, a certain genealogical irony in this failure of particular biometric systems to be able to differentiate "Asian" fingerprints. As I discussed in Chapter 1, the use of fingerprint biometrics was developed by the British in colonial India precisely in order to differentiate the homogeneous-looking natives. As Cole (2002, 127) remarks, "In the United States, as in India, fingerprint identification developed in response to the problem of identifying Asian individuals whose faces were supposedly too similar for officials to distinguish. The myth of racial homogeneity proved crucial to the cultivation of fingerprinting identification." Cole (2002, 126) quotes the *Daily Report*, at the close of the nineteenth century, celebrating the identificatory power of fingerprints: "'The thumb marks of Mon Shing, Chinese laundryman are more easily recognizable than his face'."

The failure of some Asian populations, "especially female users," to enrol in some finger-scan systems discloses the unacknowledged racialised and gendered coordinates that determine the discursive infrastructure of particular biometric systems. These racialised and gendered coordinates are clearly enunciated in Woodward et al.'s (2003, xxvii) opening position statement on biometrics: "Biometrics uses automated techniques to measure man in order to better describe himself." That the figure of "man" here is naturally assumed to be white is evidenced by the calibration to whiteness that infrastructurally marks particular biometric systems. The failure of some Asian populations, "particularly female users," to enrol in some finger-scan systems must be seen as opening up an "ontological/epistemological" split marked by gendered and racialised axes; this is a split that "pits subaltern being against elite knowing" (Spivak 1987, 268). On the one hand, this split effaces the question of racial differences from the epistemological infrastructure/software of the finger-scan technology. On the other, this effacement of "certain Asian" women from the epistemological infrastructure of these biometric technologies generates another type of occlusion and displacement. Even as certain populations of Asian women are invisibilised at this higher order level of biometric technologies, what must simultaneously remain unacknowledged is their dispatch to the empirical level of the ontological where, as labourers prized because of the very nimbleness of their "Oriental" fingers, they are instrumental in producing the very digital technologies that, at the epistemological/software level, will fail to read their fingers in the process of biometric enrolment. This failure of Asian women to be read at the operational interface of biometric systems has also been documented by van der Ploeg in the context of the U.S. Passenger Accelerated Service System (INSPASS), that "allows automated inspection of 'frequent flyers' on entering the USA." The "handgeometry required for INSPASS is biased against Asian women whose hands more often than those of others are 'too small' for the system" (van der Ploeg 2005, 125, 126).

In her analysis of the gendered and racialised dimension of the economies of the digital revolution, Suvendrini Perera (1993, 27), citing Gayatri Spivak, emphasises that "it is the urban sub-proletarian female who is the paradigmatic subject of the current configuration of the International Division of Labor." In her mapping of the geopolitical configurations of this division of labour, Perera draws attention to the neo-colonial and Orientalist selling of this urban sub-proletarian female by particular Asian governments keen "to attract technologically advanced industries" into their "Free Trade Zones." "These zones, everyone knows, offer the appeal of cheap labour combined with no taxes and almost non-existent union regulation" (Perera 1993, 26). Perera (1993, 26) elaborates on the marketing of this urban sub-proletarian female labour by quoting from a brochure issued by the Malaysian government: "The manual dexterity of the oriental female is famous the world over. Her hands are small and she works fast with extreme care. Who, therefore, could be better qualified by nature and inheritance to contribute to the efficiency of a bench assembly production line than the oriental girl?"

In her ironically titled essay, "At Your Service: Latin Women in the Global Information Network," Coco Fusco (2001, 188), in the course of her visits to the "places where the hardware of the digital revolution is assembled," identifies the process of what she terms "digital disembodiment's fiction of transcendence [that] relies on the expulsion of the abject interrelations between bodies and technologies from the virtual imaginary." Fusco (2001, 195) discloses the brutalising and expropriative dimensions of this digital "fiction of transcendence" in her mapping of Third World digital assembly plants or *maquiladoras*:

> If the Industrial Revolution created the factory town, then the digital revolution must be credited with the perfection of the export processing or free trade zone, the sites in third-world countries where low-end production takes place. In the Dominican Republic, they are nicknamed *zonas de la muerte*. Work in these territories takes place in assembly plants or *maquiladoras*, an Arabic term that entered colonial Mexico via Spain to signify the processing of foreign gains. Approximately 70 per cent of the workforce is female. In the past year, I have been conducting research on women maquiladora workers on the US-Mexico border and the Caribbean. Though these women have virtually no access to the internet, they are a crucial component of the global information circuit. Not only do they assemble much of the digital revolution's hardware, but their low wages maximize multinational profits and facilitate accelerated consumption of electronic media for the virtual class.

What both Perera and Fusco identify in their analyses of the underside of the digital revolution is the systemic construction of gendered and racialised subaltern subjects effectively put to work at the "low end" level of hardware/ontological production whilst being simultaneously excluded from the "high end" level of software/epistemological production and consumption. In the context of examining the failure of particular populations of Asian women to enrol in certain finger-scan biometric systems, the figure of the subaltern, as that figure that remains illegible within higher order epistemological systems of representation, effectively describes the non-subject status of these gendered and racialised populations.

The racialised and gendered dimensions of this epistemological/ontological split within the domain of new technologies are further elaborated in Martin Kevorkian's *Color Monitors: The Black Face of Technology in America*. In this text, Kevorkian assiduously tracks the embodied reconfigurations of this split in the context of the division of labour in U.S. information technology industries, advertisements and cinematic representations. Kevorkian (2006, 24) draws attention to the manner in which the figure of the black man is repeatedly situated, in both the workplace and in advertisements and fictional texts, as the "color monitor," hands-on-labourer, instrumentalised as a type of contemporary servile subject at the service of executive white folk: "Color monitoring, in flattering blacks' allegedly natural technical abilities, ultimately flatters the belief that being above technical work constitutes a core benefit of white privilege."

Within the material and symbolic economies of white privilege, black and coloured bodies supply the non-white, techno-ontological infrastructure that enables the ongoing reproduction of white, techno-epistemological privilege.

Critically examining the repeated representation of black men as subjects who are always at the ready to serve their white master, Kevorkian (2006, 81) brings into focus an unsettling genealogy that resonates in one particular advertisement that pictures a black man, Carl, as "naturally" embodying the figure of IT support staff behind rows of black box servers; the quip beneath this picture reads: "There are thousands where Carl came from." Kevorkian (2006, 81) questions: "And where exactly would that be? What place can be counted upon to provide an inexhaustible supply of server men? The familiar image of bodies stacked radially in the hold of a ship comes to mind as a historical echo of such language. At present, a more likely source than the African continent would be India—or from deep in the heart of darkest America."

The contemporary resonances with past histories of slavery assume even more chilling dimensions when Kevorkian (2006, 83) exposes the manner in which many IT industries are relocating their production and repair of computers to privately-run prisons that effectively exploit the African American inmates in terms of a slave-labour force stripped of any workers' rights: "The American bodies whose technological labor these companies have secured belong to the well-known demographics of the U.S. prison systems. The end of 2000 found 1.3 million people in state and federal penitentiaries; of this total, 420,000 were black men between the ages of 20 and 29." Angela Davis (1998), in the context of what she terms the "prison industrial complex," has also unpacked the racialised dimensions of the exploitation of black prison labour in the U.S., drawing attention to the systemic nature of racialised punishment and its imbrication with expropriative, multinational correctional industries. In Kevorkian's analysis of IT advertisements that tout the use of prison labour, all the bodies that are represented are black. Many of the IT advertisements and articles underline the "added value" of this captive labour force: "Men serving life sentences form the core of tech support because they stick around longer" (cited in Kevorkian 2006, 85). As Kevorkian (2006, 85) sardonically remarks,

> The tautology of this matter-of-fact observation gives new meaning to such promotions as "FREE Lifetime Tech Support." That particular long-running offer from MicroWarehouse, by the way, always appears next to the thumbnail photo of a smiling black man and an Asian woman wearing headsets. These are the simulacra that the United States wants from its tech support ... someone who has no choice but pleasantly to oblige the customer, someone who isn't going anywhere soon, someone whose time is not his own.

Biometrics' Colonial Genealogies

The colonial dimensions of this finger-scan biometric discrimination need to be historicized in more detail. As I discussed in Chapter 1, inscribed in finger-scan systems of identification and verification is a colonial moment of foundation—and I am referring here specifically to the fact that traditional fingerprint identification systems, based on acquiring a subject's inked fingerprint impression and then classifying it within a taxonomic filing system, was first developed by the colonial British administration in India "in response to the problem of administering a vast empire with a small corps of civil servants outnumbered by hostile natives" (Cole 2002, 63). In effect, the technology of fingerprint biometrics was developed in order to construct a biopolitical system of colonial identification and surveillance of subject populations in the face of British administrators who "could not tell one Indian from another" (Cole 2002, 64). Inscribed in the incipient moment of fingerprint biometrics' development is a racialised agenda driving the system's mode of identification and the biopolitical uses to which it can be put.

The colonial reach of this biometric technology extended to all the other British colonies, including Australia. In the Australian context, the colonial application of fingerprint biometrics is nowhere more graphically and movingly dramatised than in the *Cubillo* case, in which Peter Gunner, a survivor of the Stolen Generations, attempted to achieve justice for the life-long traumas that ensued once he was removed from his mother by the state. In arguing that he had been unlawfully removed from his Aboriginal mother and subsequently institutionalised, the government presented the thumbprint of Gunner's mother, arguing that the thumbprint, "on the balance of probabilities," signified "her express and 'informed consent' to her son's removal and subsequent institutionalisation" (Cunneen 2005, 70. For an extended analysis of this case and its biopolitical use of the thumbprint biometric, see Neville 2006). In his analysis of the violent asymmetries of power that were mobilised in this case, Chris Cunneen (2005, 70) asks:

> The body of the colonised becomes the site of colonial record-keeping, but what meaning can we attach to a thumbprint? During the proceedings, there could be no real examination of, or challenge to, the consent of Peter Gunner's mother in the absence of evidence. By the time of the hearing, the mother was dead and there was no way of identifying the officer from the Native Affairs Branch who obtained the thumbprint—or indeed, if the thumbprint truly belonged to her.

I want to interlace, at this juncture, two different historical moments with two related biometric technologies: British colonial India and the digital assembly plants of contemporary South East Asia, and traditional inked-fingerprint technology and digital finger-scan systems. Whereas, in colonial

British India, the "problem" of governance pivots on the white ruler's inability to discern or read for ethnic difference, a taxonomic system of fingerprint identification is developed and deployed in order to help these officials discriminate among a dangerous, because undifferentiated biopolitical mass. The consequent passing of the *Criminal Tribes Act* (1871) demanded the "registration, surveillance, and control of certain criminal tribes" (Cole 2002, 67). These certain criminal tribes were already cast within the criminalising dragnet of a form of proto-racial profiling, as the *Criminal Tribes Act* "made it possible to proclaim entire social groups criminal, on the basis of their ostensibly inherent criminality" (Cole 2002, 67). I need hardly remark here on the analogy between the nineteenth-century *Criminal Tribes Act* and the contemporary U.S. *Homeland Security Act*, which has ostensibly criminalised whole swathes of people of Arabic and/or Muslim ancestry and of people marked by the spurious ethnic descriptor "of Middle Eastern appearance" (Pugliese 2003; Parenti 2003, 204–6).

In the annals of colonial India, British officials repeatedly complain of the "problem of *racial homogeneity*": "One official complained that the 'uniformity in the colour of hair, eyes, and complexion of the Indian races renders identification far from easy, and the difficulty of recording the description of an individual, so that he may be afterwards recognised, is very great'" (Cole 2002, 67). Viewed from the contemporary context, even as finger-scan is deployed as a system of identification and verification of both civil and criminal subjects, that foundational gendered and racialised blindness to the differential contours of other bodies inflects the technology's current deployment. Put succinctly, traditional fingerprint technology's transmutation into a digital finger-scan system is marked by a graphic reversal of a discursive relation that still remains unbroken, even as it has been technologically reinscribed. Exemplified here is a structural irony that ensures that the inability of British colonial officials to read for ethnic difference has now been encoded in the systemic failure of some contemporary finger-scan systems to read for ethnic-gendered difference. Regardless of this irony, the end result is the same: certain population groups are discriminated against because of their "racial homogeneity" and their epidermal chromatism/difference.

Pixelating Race: The Racialised Colour Spectrum of Digital Discrimination

The fact that certain non-white subjects "may stand in front of a facial-scan system and simply not be found" (Nanavati et al. 2002, 66) situates these individuals within a non-locus (they are in no place even as they occupy a particular spatio-temporal location) constituted, digitally, by a zero degree of non-representation. This zero degree of non-representation operates at both literal and symbolic levels: literally, despite the corporeal presence of a subject, there appears no subject as far as the facial-scan system is concerned; symbolically, the non-white subject is reduced, within the binary logic of biometric algorithms, to a negative: in pixel values, blackness is, predictably, equivalent to the negative degree of zero, whereas whiteness is inscribed with the positive value of one ("Pixels with value one and zero are called white and black respectively" [Roddy and Stosz 1999, 41]).

Viewed in this context, non-white subjects are marked by a failure of the very animating logic of the visual image as they fail to possess, in Roland Barthes' (1993, 88–89) terms, any "indexicality" or "evidential force"; rather, they do not even appear in their appearance before the facial-scan device and therefore they are neither present nor can they be represented. And I am aware that I am deploying here, in the context of digital imaging technologies, a seemingly anachronistic term in drawing upon "indexicality"—as a term exclusively used to describe pre-digital, analogue imaging technologies (for example, analogue photography traces the light that emanates from an object onto a chemically treated film, generating an analogous or indexical presentation). Yet, I would argue that despite the fact that digital imaging constructs images out of numerical codes that may have no indexical or causal relation to the photographed object, the very logic of biometric technologies designed to identify or verify subjects is still predicated, in theory if not software practice, on establishing a type of analogous or indexical, and thus *evidentiary*, relation between the subject who is scanned and her or his stored template.

This racialised FTE inflects not only established biometric technologies such as finger- and facial-scan systems, but newer biometric systems such as iris-scan. Iris-scan systems are designed to create templates based on the imaging of a subject's iris. Yet, as Nanavati et al. (2002, 80) explain, "Locating the iris-pupil border can be challenging for users with very dark eyes, as there may be very little difference in color as rendered in the technology's 8-bit grayscale imaging." I would argue that there is encoded in the phrase "users with very dark eyes" a racialised group of non-white subjects. This is made evident when this phrase is juxtaposed against scientific literature dedicated to isolating, within the domain of forensic pathology for example, distinct "racial" characteristics for the sake of body identification: "The eye colour is useful in the caucasian race (negroid and mongoloid races virtually all have brown irises)" (Knight 1997, 32). The "problem" that some subjects' irises may have "very little difference in color" to help with the task of differentiation resounds with all the force of that caucacentric cliché that laments that black and Asian faces look all alike as they seem to possess very few congenital markers of difference. Once again, the biopolitical "problem of racial homogeneity," and the seeming "uniformity in the colour of hair, eyes, and complexion of the Indian races" (Cole 2002, 71), is actively at work here.

Nanavati et al. (2002, 37) proceed to elaborate on one of the "potential limitations of iris-scan technologies": "the ability to locate distinctive features in very dark irises. Iris-scan technology is based on 8-bit grayscale image capture, which allows for 256 shades of gray; features from very dark irises may be clustered at one end of the spectrum." This "one end of the spectrum" is analogous to what Thandeka (2000, 24), in another context altogether, terms the "non-white zone." Transposed to the context of biometric technologies, the non-white zone is constituted by a racialised colour spectrum that falls directly outside a digital grayscale image-capture system calibrated to whiteness. Within the parameters of this system, very dark irises literally fall beyond the grayscale pale. What becomes apparent, when discussing the ongoing racialised infrastructural dimensions of these technologies, is the manner in which they so clearly reproduce the binarised racial zones of the larger white culture within which they operate. I have already drawn attention to the ontological/epistemological racial split that these new digital technologies reproduce. I want to elaborate on this split by resignifying Thandeka's concept of the non-white zone in the context of biometric technologies.

Biometric technologies such as finger-scan, facial-scan and iris-scan are to be found throughout institutions and businesses primarily concerned with securing physical and/or symbolic access to important sites and/or information. As such, biometric technologies have the power to determine who may or may not enter and access critical sites of knowledge/power. The calibration to whiteness that inscribes the infrastructure of some of these biometric technologies functions, when situated in this context, to reproduce the type of stratified biopolitical zones of racialised exclusion that continue to pervade such multi-ethnic, multi-racial nations as the U.S., Australia and the U.K.

At the physical level, the non-white zones that Thandeka draws attention to are the very urban geographies whose parameters are constituted by unwritten but palpable racialised lines. At the symbolic level, that is, at the level of a culture's systems of signification, the non-white zone is determined by the systemic exclusion of non-white faces and bodies from the broad realm of representation, including film, television, advertising and so on, unless, of course, they serve to reproduce the dominant culture's stereotypes of its racialised others (see, for example, Tate 2003; Negra 2001; Rony 1998; hooks 1992). The calibration to whiteness that marks the racialised infrastructure of the biometric technologies that I have been discussing must be seen as coextensive with these racialised economies and geographies of either exclusion or inclusion. This calibration to whiteness in biometric technologies must be seen as performing a type of digital segregation and "social sorting" (Lyon 2005) of racialised population groups. Unsurprisingly, these racialised economies and geographies and their white/non-white zones perfectly articulate the struggle over the control of what Foucault and Said term the knowledge/power nexus, where control of knowledge is indivisibly linked to the maintenance of a flexible positional superiority over one's subalternised subjects.

Biometric Synecdoches and the Legal Category of the Subject

When theorised in terms of their constitutive rhetoricity, the biometric production of templates reproduces the figural logic of synecdoches: every biometric template functions synecdochically in terms of a part signifying the larger whole of the enrolled subject. I want to elaborate on the juridico-political dimensions of theorising biometric templates in terms of synecdoches by arguing that biometric templates must be viewed as also synecdoches of the legal category of the subject. In particular, the biometric template must be conceptualised as a juridico-political entity as it synecdochically functions to constitute and reproduce the legal category of the person-as-subject. In other words, when confronted by a biometric system, unless one is able to produce a template, one is directly denied the subject status of legal personhood; whether or not a subject is enabled to take up this position directly determines whether or not they may be given legal or authorised access to restricted space and/or information. In this biometric schema, not to produce a template is equivalent to having no legal ontology, to being a non-being.

The indissociable relation between the question of representation and the very possibility of political agency in and through law is brought into sharp focus in Peter Goodrich's (1990, 263) naming of this relation as one that is played out in "the theatre of attachment." Through his invocation of this reflexively representational metaphor, Goodrich (1990, 263), drawing on the work of Pierre Legendre, delineates the manner in which the legal subject is predicated on the "forms of attachment to law":

> The question is initially that of the theatre of attachment, an issue of the mask or role or identity that will bind the individual to law, that will tie together in legal form the unity of a lived existence and so secure the political agency of the human subject through representation.

My reading of the biometric template in terms of the figure of synecdoche (as instrumental in constructing the legal category of the subject) must be seen as a contemporary, *in silico* version of what Goodrich (1990, 263–64) terms an "actor's mask":

> Note that one of the central constructions of civil law, that which, following Justinian's terminology, we call the *law of persons*, literally derives from *persona*—referring initially to an actor's mask—and authorises me to translate the formula *de iure personarum* by "of the law of masks." In all institutional systems the political subject is reproduced through masks. This translation contributes to the rehabilitation of the problematic of the image at the heart of the legal order.

What Legendre and Goodrich disclose, in their critical genealogies of law, is the systemic erasure of the image at the very heart of law even as it establishes the (disavowed) conditions of possibility of the legal subject as representational being, as actor and agent in the socio-legal theatre. In articulating the importance of zero in Legendre's genealogies of law, Goodrich (1990, 281) underscores its value, in law, as "a lack, an absence that is filled by entry into the symbolic." If, as I argued earlier, the non-white subject is often dispatched to a zero degree of non-representation by biometric technologies calibrated to whiteness, then she or he is precluded from entry into the agentic domain of the symbolic "where the ego [as notationally equivalent to zero] becomes person, a subject of law": "The entry of the individual into the symbolic, the transition from zero to one, from lack to identity, is the condition of institutional existence, the capture of the subject by law. What is at stake in the order of reference and in the 'name of the law' is precisely the possibility of social speech—and there is no other speech—and so the possibility of being human, or in scholastic terms of becoming 'a speaking being'" (Goodrich 1990, 281, 282). The impossibility of occupying the position of "a speaking being," "of being human," inserts biometric FTE into the domain of the biopolitical and its attendant effects. Biopolitics, Foucault (2003, 256) notes, is not only marked by the "right to kill," in the sense of literal murder; it must also be understood "quite simply [as producing] political death, expulsion, rejection, and so on" (see also Agamben 1998). I am not arguing here that biometric FTE is a form of conscious state-sponsored conspiracy openly directed at people of colour; rather, in deploying the term *infrastructural whiteness*, I am pointing to the operation of institutional racism as something that is embedded within the very infrastructures (its technologies, institutions, apparatuses) of the societies in question, consequently enabling various forms of political death or non-representation.

The transition from zero to one, from absence to representational subject-as-legal-person, is precisely what certain biometric systems calibrated to whiteness foreclose when faced with non-white subjects. Inscribed in this failure of transition is also a failure of ethics. Understood in Levinasian terms, this literal biometric effacement of the face of the other produces a failure of ethics because it obliterates the possibility of justice in the context of embodied differences: "There must be a justice among incomparable ones. There must be a comparison between incomparables and a synopsis, a togetherness and contemporaneousness; there must be thematization, thought, history and inscription. But being must be understood on the basis of *being's other*" (Levinas 1991, 16). I cite this passage not only to bring into focus the failure of ethics in the face of the symbolic violence that biometrically occludes the face of the other, but also in order to complicate facile and reductive understandings of the operations of ontology in Emmanuel Levinas' work in relation to questions of embodied difference and the material conditions of achieving justice. "The way of thinking proposed here," writes Levinas (1991, 16), "does not fail to recognize being or treat it, ridiculously and pretentiously, with disdain, as the fall from a higher order or disorder. On the contrary, it is on the basis of proximity that being takes on its just meaning." Proximity, in Levinas, marks ethics in terms of a relation that does not efface or occlude the face of the other; rather, it opens the possibility for inscription between incomparables so that thematisation and ontology can only be thought, ethically, "on the basis of *being's other*."

In the context of everyday life, the biopolitical segregation enabled through FTE divides into the three fundamental categories that constitute the practical application of biometric systems—it can preclude a subject in terms of gaining "*logical access* to data or information"; "*physical access* to tangible materials or controlled areas"; as well as identifying or verifying "the identity of an individual from a database or token" (Nanavati et al. 2002, 144; italics original). The racialised practices of segregation that I have been discussing dovetail perfectly with what Lyon (2003, 81), in his analysis of surveillance post-September 11, terms "digital discrimination," which "consists of the ways in which the flows of personal data—abstracted information—are sifted and channelled in the process of risk assessment, to privilege some and disadvantage others, to accept some as legitimately present and reject others," and this is "increasingly done *in advance* of any offence."

It is in this context of the hyper-surveillance of targeted racialised subjects that I would argue against naïve celebrations of the type of failure of biometric representation that I have addressed. At a recent conference, for instance, after the delivery of a version of this chapter, a number of responses from the audience argued that it was a good thing that certain subjects seemingly escaped the imaging capabilities of some biometric technologies. Biometric non-capture of a subject's image was viewed as a type of positive loophole or escape clause from contemporary systems of identification and surveillance. As I have attempted to demonstrate, this view entirely disregards the critical question of equity of access within civic spaces/institutions and the manner in which FTE systemically precludes particular racialised subjects from accessing both physical sites and knowledge/power. Moreover, I term this celebration of failure of representation as "naïve" because it is predicated on a liberal-humanist understanding of contemporary systems of surveillance, where the question as to whether one gets visually imaged or not remains merely a question of choice. I would argue, on the contrary, that visual surveillance, in the context of omnipresent CCTV, ATMs and other related technologies, is always already at work.

Indeed, even when such imaging technologies as CCTV surveillance cameras that are calibrated to whiteness fail to capture a clear image of a non-white subject, the mere black "blob" of an image that they do capture can still be mobilised in the court of law to indict non-white subjects for alleged criminal offences. And I am referring here to the *R v Mundarra Smith* case (97/11/0742), where a series of still images derived from a National Bank of Australia CCTV camera were used to indict a young Aboriginal man, Mundarra Smith, for bank robbery. Despite the fact that the photographs literally showed nothing more than an unidentifiable black figure, "the Crown called two police officers from the Redfern patrol who gave evidence that they knew Mundarra Smith from their policing duties and that, when they looked at the photographs, they recognised him in them" (Biber 2005). As Katherine Biber (2005) writes in her analysis of this case, the case finally went to the High Court, where "The jury was given additional directions about the dangers of making 'cross-racial identifications.' The jury acquitted Smith who had, by the time of his acquittal, served 3 years and 6 months of his original sentence."

I want to conclude this section by drawing attention to recent interventions by Chinese and Japanese biometric scientists working to construct biometric systems that would in fact be calibrated to capture people of colour whose features might otherwise fail to be biometrically read. Li Feng, Jianhuang Lai and Lei Zhang (2004, 268), for example, are working on developing new algorithmic biometric formulae designed to be responsive to what they term "the complex and individual shape of the face and the subtle and spatially varying reflectance properties of the skin." As I discussed earlier, in the biometric literature, talk of the "reflectance properties of the skin" must also be seen as encoding a type of epidermal chromatics of race. In the Japanese context, Shihong Lao and Masato Kawade (2004, 344) are in the process of developing what they term "ethnicity estimation" for biometric facial feature extraction. What is interesting about this work is that it signals an attempt reflexively to integrate racial and ethnic differences into the operational software of biometric systems, and thus override homogenising white templates. However, in the Lao and Masato (2004, 345) text, the question of ethnicity is reduced to three categories: "European, Asian, African." Again, as I discussed earlier, situated within the genealogy of racialising and racist discourses, these three taxonomic categories are at once both hierarchical and, because of their globalising essentialism and homogenisation, not necessarily conducive to reading the complex nuances of embodied ethnic differences.

Such moments in the biometric literature, where Eurocentric racial categories, taxonomies, values and presuppositions are unreflexively reproduced by computational scientists of colour, compel me to begin to account for the discursive forces at work in the very reproduction of Eurocentric systems of value by non-western subjects. I would argue that one could begin to account for the reproduction of such Eurocentric values and categories by proceeding to view science, technology and bodies as always already socioculturally inscribed and positioned. Contrary to the sorts of statements articulated within the biometric literature, that position these new technologies as objective, "human-free" and not "reliant on racial stereotypes," technologies, and bodies, are always influenced by the complex interplay of historical, political and sociocultural factors that determine both their emergence and the uses to which they are put. Biometrics, as technologies predicated on the mathematical measurement of bodies, emerged, as

I discussed in Chapter 1, out of the colonial-scientific practices of anthropometry, craniology, phrenology and so on. As has been well established by a considerable body of scholarship, these colonial-scientific practices were fundamentally oriented by racial categories and hierarchies, with whiteness located as the apex of the hierarchy.

Viewed in Foucauldian terms, then, the knowledge/power nexus that inscribes biometrics is already oriented by what Ferdinand Henrique terms "white bias"; Kobena Mercer (1990, 250) elaborates Henrique's term by examining "the way ethnicities are valorised according to the tilt of whiteness." In the context of a scientific field dominated by the U.S., Europe and the U.K. (in terms of research centres and academic publishing), for biometric scientists of colour who reproduce in their work Eurocentric values, "European elements ... [are] positively valorised as attributes enabling upward mobility" (Mercer 1990, 250). Moreover, I would argue that these relations of racialised power, precisely because they are discursive, be viewed as establishing fundamental conditions of "intelligibility" for the very broaching of questions of race and ethnicity. Thus, for example, even as one would assume that a Chinese, Indian or Japanese scientist would be well aware of the problematics and differences that inscribe an essentialised category like "Asian," the very intelligibility of a discussion of the category of ethnicity has to be cast into a homogenising category that is intelligible to the west. The rewards—financial, social and political—for reproducing "white bias" have been extensively documented (see, for example, Frankenberg 1993; Thandeka 2000; Yancy 2005).

"Being Invisible and Without Substance"

> "Being invisible and without substance, a disembodied voice ... what else could I do? What else but try to tell you what was really happening when your eyes were looking through?"
>
> (Ellison 1972, 439)

I want to conclude this chapter by underscoring the continuing failure of biometrics critically to theorise the relation between the body and technology. Perhaps what is most striking about this critical failure is that the entire logic of biometric technologies is predicated on *the body as "the password for access"* (Edwin P. Rood cited in Woodward et al. 2003, i. Emphasis added). It is here that I locate the crux of what is problematic about these biometric conceptualisations of the relation between the body and technology. Within the conceptual schema of the biometric sciences, the body is represented in terms of a purely biological datum that is then processed by particular technologies designed to acquire a digital simulacrum or algorithmic template of its distinctive features. What remains untheorised here are two key points. First, the body that presents itself before a biometric system is not a purely biological datum; on the contrary, as discussed in the Introduction, it is already somatechnically inscribed with a system of technological mediations and signifying relations that mark it *before the fact* of physical presentation. That the body is already somatechnically mediated before the fact of biometric presentation becomes graphically evident in the face of the digital discrimination that is enacted by certain biometric technologies when particular non-white subjects present themselves for enrolment. Before the fact of attempted biometric enrolment, the body has already been marked and made legible or not according to both its race and gender. To the categories of race and gender can be added age and class (and, as I discuss in Chapter 3, disability), as often a person's age will affect whether or not, for example, a fingerprint is legible to a biometric system due to the increasing corrugations of the skin because of the aging process; similarly, the fingerprints of manual labourers are often disfigured and thus illegible because of the abrasions, cuts and bruises they experience in the course of their work (Jain and Ross 2008, 16).

Second, biometric technologies are not the neutral, objective systems that their proponents argue they are. On the contrary, their seemingly pure geometry, their transcendental algorithms and apparently unmotivated digital formulae are all inflected with the dense sociocultural and historical significations of their designers. The subject's presentation for enrolment before a biometric system, then, exemplifies a moment where a dense network of mediations is already at work. This is not to say that the corporeality of the flesh is somehow made less relevant because of these mediations. Rather, the point of interface between these biometric systems and the mediated body is what precludes the mystical transcending of the irreducible corporeality of the body—as always already somatechnically signifying corpus. And I mark this point in the face of repeated claims celebrating the increasing irrelevance of the corporeality of the body (for example, as epidermal figure) in the context of new information technologies. For instance, in his discussion of "the effects of technological acceleration arising from digital processing and computer-mediated communications," Paul Gilroy (2004, 108) argues that these effects "mean that the individual is even less constrained by the immediate forms of physical presence established by the body. The boundaries of the self need no longer terminate at the threshold of the skin." It is, on the contrary, the threshold of the skin that continues, in the context of biometric technologies, to constrain the physical presence established by the body. The chromatic contours of the skin, its very epidermal (ir)reflectivity in the face of technologies calibrated to an infrastructural whiteness, determine whether a subject will be read as present or not in the process of biometric enrolment.

If the racialised contours of the epidermis do not terminate at the threshold of the skin, it is because they extend beyond the physical parameters of the subject into the larger biopolitical domain, inflecting the operating functions of technologies and determining critical questions of knowledge/power, equity and access. At the threshold of the skin, that empirical point of terminus dividing the corporeal from the non-corporeal, the *bios* of the skin is metaphorised into its seeming other: *technè*. And, in a reverse direction, *technè* is metaphorised as body: literally, *biometrics*. The plasticity of integument, the chromatics of the dermis, the phenotypicality of its racialised morphology, its epidermal reflectivity—all these aspects of skin overflow the threshold of the embodied subject as they are assimilated into the *technè* of culture. All these features of flesh, through metaphorical exposition and extension, constitute a continuum across the threshold of embodied subject and technological object. Simultaneously, algorithms, optical settings, lighting—all these technological features function to determine the very possibility of the subject to appear as embodied template figure. Biometrics dreams of a univocal deployment of pure geometry in its desire objectively to apprehend the subject of flesh. Yet already encoded in the machine, already orienting its algorithms and pixelated templates, is a phantasmatic body that it fails to acknowledge: invisibilised because white, this body must remain unthought even as it overdetermines the very possibility of biometric appearance/non-appearance.

3 "Identity Dominance": Biometrics, Biosurveillance, Terrorism And War

Biometrics as Technologies of War

In his tracking of the emergence of biopolitics as a technology of power, Foucault (2003, 246) draws attention to the manner in which regimes of biopower are driven by a need for "regularisation" that makes itself manifest in the deployment of "forecasts" and mechanisms of "security." In a post-9/11 context, configured by such events as the war on terror and the U.S.-led war in Iraq, biometric technologies have been deployed both as "forecast" systems that can purportedly detect, in advance of the fact of someone actually committing a criminal act, a suspect figure with criminal or terroristic intent, and as "securitising" technologies that will ensure, in Foucault's titular phrase, that "society will be defended." In this chapter, I examine the manner in which essentialised biotypologies are being mobilised and reproduced within biometric systems in order, pre-emptively, to identify, surveil and capture targeted subjects. Biotypologies, I argue, function to constitute targeted subjects in terms of clusters of essentialised, and visibly identifiable and contained, corporeal features, behaviours and practices. Genealogically tied to such seemingly outdated disciplines as anthropometry, phrenology and criminal anthropology, the use of biotypologies by both the military and law enforcement agencies reproduces a disciplinary biopolitical regime premised on normative conceptualisations of race, gender, (dis)ability and bodily behaviour.

In their Introduction to the field of biometrics, Woodward et al. (2003, xxiv–xxv) write:

> Following the September 11, 2001 terrorist attacks, the U.S. government and other governments and organizations throughout the world became greatly interested in this emerging [biometric] human recognition system ... Biometrics are an integral and distinctive part of human beings. As such, they offer a natural convenience and technical efficiency that other authentication mechanisms, which must be mentally remembered or physically produced, do not. For this reason, biometrics can provide identity assurance for countless everyday activities currently

protected by traditional means of access control—cards, personal identification numbers (PINs), and passwords.

Woodward et al. open their Introduction by invoking the spectre of that watershed date, 9/11, in order to mark the inextricable relation between the war on terror and the consequent deployment of biometric technologies by government and military institutions in this waging of war. In this chapter, I situate biometric technologies within the context of war understood in its broader biopolitical sense. "War," Foucault (2003, 51) argues, "is the motor behind institutions and order. In the smallest of cogs, peace is waging a secret war. To put it another way, we have to interpret the war that is going on beneath peace; peace is a coded war." In examining the techniques and apparatuses deployed by the state in securing peace, I am interested in disclosing how they in fact operate as technologies instrumental in the conduct of war. War, in biopolitical terms, refers not only to the literal conduct of fighting in a theatre of war; it simultaneously refers to the state's "polymorphous techniques of subjugation" (Foucault 2003, 27). These polymorphous techniques of subjugation are being deployed by the state against biosocial "problem" populations, including refugees and asylum seekers, Roma people, Iraqi "insurgents" and so on. Situated in this context, I draw attention to the powerful convergence of technologies of biosurveillance (such as networked multimodal biometric databases) and new and emerging biometric systems of identification and verification such as gait signature biometrics and Project Hostile Intent. This convergence of biometric and surveillance technologies must be seen as generating a biopolitical theatre of war constituted by regimes of biosurveillance targeting racially profiled subjects who, by definition, are situated as interlopers within the body politic of the nation and who risk being criminalised and imprisoned before the fact of having committed any criminal offence. In what follows, I proceed to examine the biopolitical ramifications of these biometric technologies.

"Frontier Technologies" in the Service of Unmasking Terrorist Stealth-Masking: Biometric Gait Signature and the Rhetorics of Walking

In 2005, the International Association of Chiefs of Police (IACP) published a document, "Training Keys #581: Suicide (Homicide) Bombers: Part 1 [TK #581]," designed to assist law enforcement authorities in the capture of prospective suicide bombers. The IACP is an influential law enforcement organisation that has an international reach. Established in 1893, it claims on its website to have over 20,000 members in over 89 different countries. Its influence in shaping law enforcement policy is also documented on its website, where it claims to have spearheaded the adoption of "breakthrough technologies and philosophies" that have revolutionised police practice (IACP).

I map this background to the IACP in order to underscore its power and influence in shaping attitudes and policies in the domain of law enforcement.

TK #581 announces in its Introduction that its primary objective is to offer the reader "profiles" of suicide bombers in order to enable law enforcement personnel to prevent attacks. In the first instance, suicide bombers are scripted as at once graphically anomalous in the context of normative culture and yet invisible. This paradoxical feature is what gives them inordinate power: grossly aberrant in their anti-social values, yet they appear to pass through social spaces without detection. Indeed, this feature is what enables a suicide bomber to function "as a precision weapon, taking the explosive device right to the target. This is a dimensional stand-off attack in the sense that the terrorist is 'invisible' (stealth-masked) until the device is detonated, which helps overcome the Western advantage of stand-off targeting and defense based on physical distance" (IACP 2005, 2). It is precisely the doubleness of suicide bombers that disturbs, generates anxiety and foments panic: monstrously deviant yet they are invisible. Kelly Gates (2006, 418) has fleshed out this paradox in the context of the post-9/11 visual representation of Arab men as "simultaneously 'unidentifiable' and readily identified by their characteristic 'faces of terror'." In this racialised context, Gates proceeds to examine the mobilisation of biometric facial recognition technologies in pursuit of these "fetishized objects."

In the context of the ruses and impersonations that a prospective terrorist can deploy in order to dupe human systems of surveillance and identification, so-called "frontier technologies" (a term I will presently examine in some detail) are being mobilised by law enforcement and military organisations in order pre-emptively to capture terrorists. The use of biometric systems as technologies mobilised to capture suspect subjects in advance of their having committed any offence is clearly evidenced in two emergent biometric technologies funded by the U.S. Department of Defense's Advanced Research Projects Agency (DARPA): the Video Early Warning (VEW) system and the Human Identification at a Distance (HID) project. VEW and HID are emergent biometric technologies that will screen a subject's behavioural features, specifically, the modality of their gait, that is, the way they walk.

> "When gait research is perfected, its uses for anti-terrorism surveillance will be invaluable," says a former program manager at DARPA who asked that his name not be used. He mentions the possibility of having software that would detect whether a stranger walking into a facility is a frightened woman or a terrorist with hidden explosives.
>
> (Geracimos 2004)

In his essay "Walking in the City," Michel de Certeau (1988, 97) stages a critical re-evaluation of walking by reading the kinesiology of a subject's walk in terms of rhetorics. In the process, he identifies a number of "enunciative functions" for what he terms "pedestrian speech acts." These enunciative functions, amongst other things, serve to make intelligible the seemingly random practice of walking in terms of cultural rhetorics to which can be attributed a series of values, including a "truth value," an "epistemological value," and an "ethical or legal value." It is on this last value that I want to focus in order to begin critically to address the cultural politics of gait signature within a biometric counter-terrorist network. The enunciative modality of a subject's walk in terms of its ethical or legal value refers to "'deontic' modalities of the obligatory, the forbidden, the permitted, or the optional. Walking affirms, suspects, tries out, transgresses, respects, etc., the trajectories 'it speaks'" (de Certeau 1988, 99).

The biometric conceptualisation of a subject's walk as constitutive of an identificatory *signature* perfectly dovetails with de Certeau's theorisation of a subject's walk as ineluctably inscribed by a rhetorical systematicity. In the language of biometrics, the rhetorical systematicity of a gait signature is "characterized by the joint angles between body segments and their relationships to the events of the gait cycle" (Yoo, Nixon and Harris 2002). In biometrics, however, this rhetorical periodicity of human gait motion is predicated on a set of algorithms entirely removed from the sociocultural contexts within which all subjects are situated and within which the biometric algorithms are themselves constructed. In contrast, de Certeau's approach insists precisely on situating patterns of human locomotion within the dense sociocultural fabric that inflects a subject's kinematic characteristics.

Viewed in this light, biometric uses of gait signature for counter-terrorist practices must be seen as conceptually situated within the domain of *deontology*, as locus of the ethical, the legal, the transgressive and the criminal. The tracking of suspect subjects will rely on the involuntary disclosure of the unique modality of their walk. Their walk signs or articulates the letters of their "name" with every ambulatory step they take:

> And the target doesn't have to be doing a Michael Jackson moonwalk to be distinctive because the radar detects small frequency shifts in the reflected signal off legs, arms and the torso in a combination of different speeds and directions. "There's a signature that's somewhat unique to the individual," says [Gene] Greneker [Georgia Institute of Technology, Defense Advanced Research Projects Agency]. "We've demonstrated proof of this."
>
> (Sniffen 2003)

The gait signature of the target subject will disclose whether they are listed on the Pentagon's Total Information Awareness database:

> a vast surveillance system ... Conceived and managed by retired [Iran-Contragate's] Adm. John Pointdexter ... TIA is an effort to design breakthrough software "for treating these databases as a virtual,

> centralized grand database" capable of being quickly mined by counterintelligence officers even though the data will be held in many places, many languages and many formats.
>
> (Sniffen 2003)

Situated within the domain of the deontological, the biometric gait signature of the suspect will determine, in advance of the fact, whether they are transgressing a security space or are in the process of committing some other crime:

> The system could be used by embassy security officers to conclude that a shadowy figure observed a few hundred feet away at night in heavy clothing on a Monday, Wednesday and Friday was the same person and should be investigated further to see if he was casing the building for an attack, Greneker said.
>
> (Sniffen 2003)

Already encoded in the trope of the "shadowy figure" is the Orientalist construct of the suspect figure "of Middle Eastern appearance" (Pugliese 2003). In other words, already inscribed within the kinesiology of the biometrics/rhetorics of gait signature is the spectre of racial profiling. Located at the very inception of kinesiology, as a discipline committed to the study of body movement, tonus and posture, is a concern with the racialised dimensions of kinesics. Writing at the beginning of the twentieth century, the anthropologist Marcello Levi Bianchini (1906, 17–18) was concerned to develop a kinesiological biotypology of race by arguing that fundamental motor and gestural differences separated the "primitive" races from more "advanced" races:

> A psychological characteristic intrinsic to primitive peoples, which still exists in Calabria, is the kinesiological equivalence of the sentiments. This law can be articulated in the following manner:
>
> In primitive societies all the affective reactions of the individual and collective psyche are essentially manifested by phenomena of movement, investing, in the process, complex and numerous motor expressions that are always inadequate to the stimulus. In all these cases, the more evolved societies have either reduced to a minimum these manifestations by substituting them with a simple word, with writing, with a measured gesture, or they have abolished them altogether by burying them in silence.

Unsurprisingly, the kinesiological biotypology of race, and its binary of "advanced" and "primitive" societies, divides, in Bianchini's (1906, 14) schema, along well-established hierarchical lines: advanced Northern

Europeans at the top and, beneath them, the primitive rest, including Africans and Arabs. In the contemporary context, this racialised kinesiological binary is graphically evidenced in western media representations of Arab peoples protesting against U.S. imperialism: massed in undifferentiated hordes, gesticulating wildly, ululating and chanting, they embody the face of Bianchini's (1906, 18) "primitive" peoples, inscribed with "hyperbolic kinesic forms" that have "excessive duration and exaggerated proportions" when compared with the "rarefied affective kinesic expression of Northern Europeans." Bianchini's kinesiological theories were influenced by the famous criminal anthropoligist, Cesare Lombroso (see Chapter 1), who, in the fourth edition of *L'Uomo Delinquente* (*Criminal Man*) (1889, 242–43), argued that "Wild gesticulation characterizes idiots, children, and demented individuals" and that, furthermore, this was also evidence of those who were "born criminals," demonstrating "a dominance of the body over the mind and the absence of psychological inhibitions over motor reflexes." Lombroso, in this same edition of *Criminal Man*, had already commenced the labour of biometrically differentiating the gaits of the three types he termed "normal," "delinquent" (thieves, assaulters and rapists) and "epileptic." Lombroso's biometrics of "normal" and "deviant" gaits was evidenced in an illustration in which the walking gaits of all three types were juxtaposed against each other (Figure 3.1).

In Figure 3.1 the gaits of the three types are illustrated through the use of footprints walking along parallel lines that diverge into different gaits; Lombroso positions, in biopolitical terms, the "normal" gait as the template against which the other two gaits emerge in all their self-evident kinesiological deviancy. The epileptic, in this context, is at once criminalised and pathologised in her or his deviation from the disciplinary regime of normative gaits: "The inexorable weight of statistics confirms the family relationship—indeed the unity—of criminality, moral insanity, and epilepsy" (Lombroso 2007 [1876–97], 257).

The biopolitical genealogy and the cultural inscription of the rhetorics of walking and gait signature that I have tracked bring into focus that other magnetised binary, Christian/Muslim, when contextualised in the contemporary "clash of civilisations" propaganda generated by the war on terror. Citing the earlier-mentioned biometric project on the identification of terrorists by their gait signature, the online McDonough Presbyterian Church states:

> Researchers at Georgia Tech, sponsored by the Defense Advanced Research Projects Agency, have reported a success rate between eighty and ninety-five percent in identifying individuals simply by their gaits, and they hope before long to increase the accuracy to the "high ninety percent range."
>
> Does your "*gait signature*" give you away as a follower of Jesus, as one who walks with the Lord?

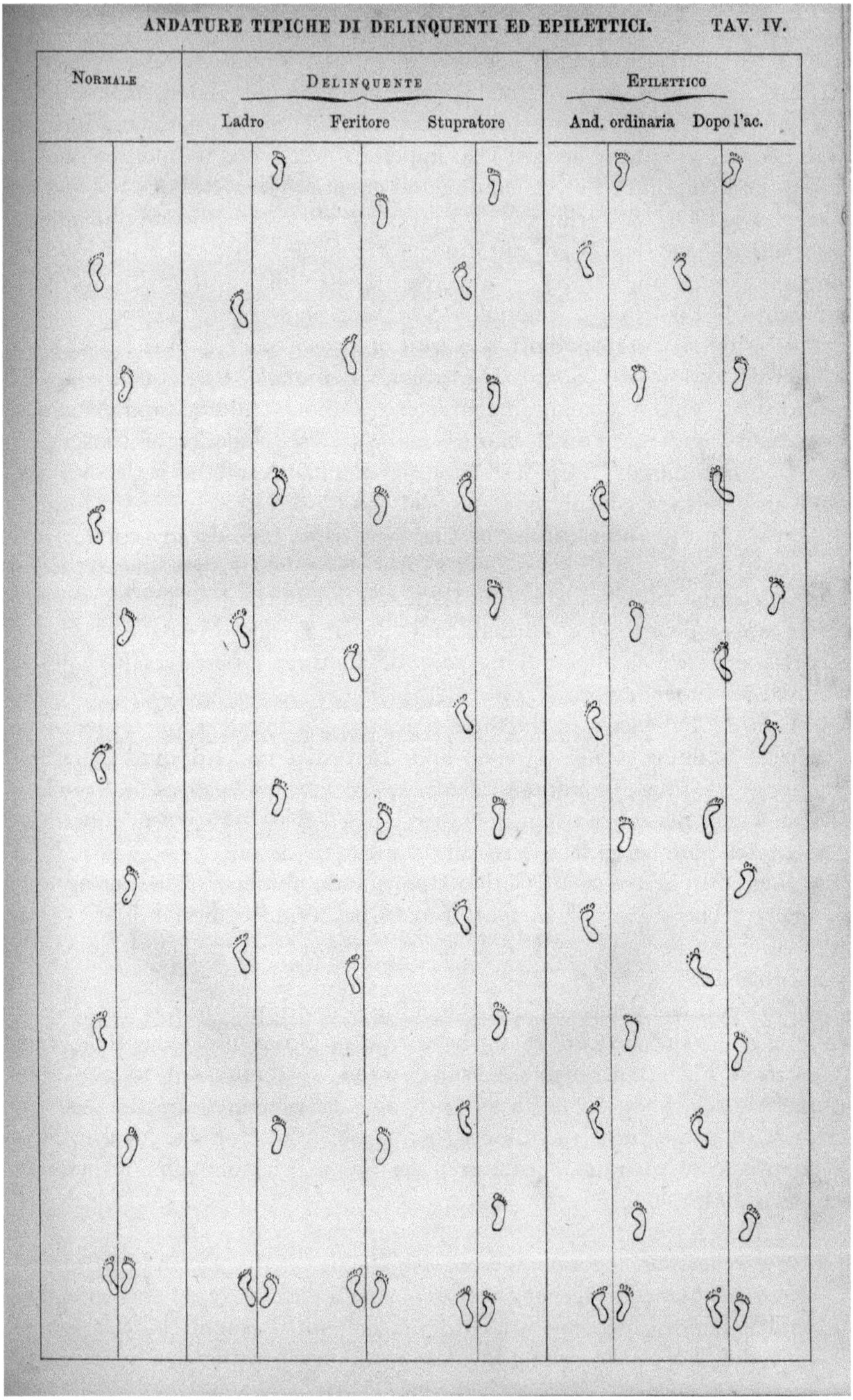

Figure 3.1 Typical gaits of normal, criminal and epileptic subjects. Cesare Lombroso, *L'Uomo Deliquente*, 1889.

> We, who claim to be Christians, are Jesus' hands and feet in the world today ... Christ's contemporary disciples charged to
>
> "Go out and train everyone you meet, far and near, in God's way of life; marking them by Baptism in the threefold name: Father, Son and Holy Spirit.
>
> By this we may be sure we are in Christ: anyone who says they abide in Christ ought to walk in the same way in which Jesus walked."
>
> *(Chorus) "And they will know we are Christians by our walk, by our walk, Yes, they'll know we are Christians by our walk."*
>
> (McDonough Presbyterian Church 2003)

The McDonough Presbyterian Church's Christian evangelising exhortation brings into sharp focus the sociocultural and political inflections that mark gait signatures. What is articulated in this text is an unsurprising point of political intersection between imperial military concerns and religious missionary agendas. The biometric of gait signature deployed to advance western counter-terrorism strategies is here neatly sutured to a missionary, religio-moral world view that remains implicit in the moral-imperial language of the former. Lieutenant Colonel Kathy De Bolt, Deputy Director of the Army Battle Laboratory at Fort Huachuca, Arizona, explains the moral aims and goals of another biometric system that is being developed in order to realise a global and omniscient capability of eyes and ears everywhere: the Biometrics Automated Toolset (BAT)—"Any place we go into—Iraq or wherever—we're going to start building a dossier on people of interest to intelligence ... We're trying to collect every biometric on every bad guy that we can" (Associated Press 2002). The BAT system is part of the U.S. Department of Defense's mission to establish "a *universally* biometrically-enabled, strongly-identified and assured, global information grid" (Office of the Under Secretary of Defense for Acquisition, Technology, and Logistics 2007, 7). The imperial capabilities of this global information grid will be enabled by the development of multi-biometric databases that are effectively networked and interoperable.

In TK #581's bullet-point profile of suicide bombers, kinesiology is seen to supply this preincident indicator: "*An unusual gait, especially a robotic walk.* This could indicate someone forcing or willing himself or herself to go through with a mission" (IACP 2005, 5; italics original). The normative kinesiological regime inscribed in this preincident indicator overlaps with the encoding of "normal" and "deviant" gaits in the operations of gait signature technology. In the *Bulletin of the Atomic Scientists*, for example, Linda Rothstein (2003) describes how gait signature technology will enable the detection of criminals or terrorists in advance of their having committed any offence: "Perhaps then we will see proud police officers, when asked how they were able to apprehend some criminal or terrorist, explain simply, 'There was something funny about the way he walked'." Inscribed in the operational logic of gait signature is the disciplinary intertwining of biotypology and anthropometry. A visually identifiable "normal" or "deviant" biotypology of gait is technologically measured through anthropometric software in order to produce an effective result. In this case, the visible biotypology of a subject's gait, measured against infrastructural normative standards, will disclose an internal moral morphology: the "funny" walk of the target subject will disclose his or her incipient criminal mind. Gait signature must, in this biopolitical context, be situated in the well-established domain of those other infamous anthropometric sciences dependent on visible corporeal indices in order disclose the internal (im)moral qualities of target subjects: phrenology, physiognomics and craniology.

The visible, preincident kinesiological indicators of suicide bombers—the so-called "robotic walk" and/or "funny" walk—are predicated on a biopolitical regime of able bodies that know of no physical or psychological impediment to walking as such. In other words, gait signature screening will inevitably catch in its net people with walking disabilities precisely because they fail to conform to a normative template walk, especially if these subjects are, in addition, marked by racially charged biotypological indices (for example, "of Middle Eastern appearance"). The disciplinary dimensions of this kinesiological regime become clearly apparent for people with walking disabilities when situated in the context of so-called "gait training" regimes that demand the kinesiological "rectification" of a disabled subject's "problematic" gait so that it conforms to a normative template (Snyder and Mitchell 2006, 94). Gait rectification technologies form part of the biopolitical apparatus of "orthopedic instruments," that is, instruments concerned with the "correction, training, and taming of the body" (Foucault 2006, 105–6). As Lennard Davis (2006, 3) writes, "the 'problem,'" in such situations, "is not the person with disabilities; the problem is the way that normalcy is constructed to create the 'problem' of the disabled person." The manner in which biometric technologies are predicated on normative schemata of the body is made evident in both the terms deployed to the describe the operating procedures of the technologies, such as "geometric normalization," and the actual infrastructural operations of the technologies: "geometric normalization" "entails the detection," in face scanning systems for example, "of two or more points on the face, and the subsequent affine mapping of the acquired image onto canonical geometry" (Socolinsky 2008, 302). "Geometric normalisation" and "canonical geometry" disclose the exercise of disciplinary power through the normative instrumentalisation of the body.

That normative conceptualisations of the body underpin contemporary biometric technologies is further evidenced in another U.S. government report on the problem of biometric Failure to Enrol: "People who cannot use voice systems, and people lacking fingers or hands from genital disease, surgery, or injury cannot use fingerprint or hand geometry systems. Although between 1 and 3 percent of the general population does not have some body part required for using any one biometric system, they are not normally counted in a system's FTER" (U.S. General Accounting Office 2002). Even while acknowledging the not insignificant percentile preclusion of people with disabilities from biometric enrolment, this government report makes clear that such subjects are not made to count in a biometric system's Failure to Enrol Rate. This effectively signifies that people with disability are doubly erased. Reproducing the same type of preclusive logic that I discussed in relation to people of colour and FTE, subjects with a disability fail to count on two critical levels: they are systemically precluded from biometric enrolment because of the manner in which the technology's image acquisition system is calibrated to a normative body; simultaneously, subjects with a disability are reduced to a zero degree of non-representation, as they literally fail to be counted as human subjects who have failed to enrol in a particular biometric system. It is in this sort of normative context that Sharon Snyder and David Mitchell (2006, 21) ask: "What can we learn about disability by beginning with the premise that our understanding of human variation has been filtered through the perspectives and research of those who locate disability on the outermost margins of human value?" In the aforementioned U.S. government report, we learn that people with disability have been biometrically dispatched to those "outermost margins of human value" where, as subalterns, they can neither represent themselves biometrically nor are they seen as possessing sufficient human value in order to be represented as such, not even at most the minimal and circumscribed of all levels, where they are merely counted in the statistics of FTER. In this context, the biopolitical ramifications of biometric technologies once again come to the fore, as they play instrumental roles in the social segregation of a human population, determining technologically and scientifically a subject's worth according to infrastructural normative standards and calibrations.

Biometric "Arms Race" and the Biometrics of "Identity Dominance" in Theatres of War

John Woodward, Jr., former director of the U.S. Department of Defense Biometrics Management Office, in his "Using Biometrics to Achieve Identity Dominance in the Global War on Terrorism," writes:

> Just as the US military has established its superiority in other arts of war, now, working with other US government organizations, it must strive for identity dominance over terrorist and national-security threats who pose harm to American lives and interests. In the context of the Global War on Terrorism (GWOT), identity dominance means US authorities could link an enemy combatant or similar national-security threat to his previously used identities and past activities, particularly as they relate to terrorism and other crimes.
>
> (2005, 30)

In the context of the GWOT, biometric technologies that are being deployed by the U.S. to help achieve the desired goal of identity dominance. Woodward (2005, 34) argues that:

> We can use biometric technology to achieve identity dominance and must deploy it to meet the requirements of force protection, actionable intelligence, and law enforcement. Establishing identity dominance through a comprehensive [biometric system] will enable the US military to identify friend or foe to keep America safer.

The importance of biometric technologies in the context of national defence is clearly outlined in the Report of the Defense Science Board Task Force on Defense Biometrics (commissioned by the Office of the Under Secretary of Defense for Acquisition, Technology, and Logistics). The Defense Science Board Task Force on Biometrics (2007, 7) writes that it

> understood its job was to examine a topic that was urgent, complex, somewhat new and distinctly open-ended. "Biometrics" was and is seen as an emerging field of growing importance to the Department of Defense and the nation's security more broadly.

The urgency of the Report is underscored by the declaration that what is unfolding in the current global context is effectively a "biometric 'arms race'" (2007, 50). The Report (2007, 7) argues that the establishment of identity dominance is predicated on biometric "identity management": "Biometric identification supports identity management, which is a key to success in many mission areas in the Department of Defense and in the larger national and international security context both in the US and internationally."

The emergence of these two terms, "identity dominance" and "identity management," enunciates, as I explain later, the assimilation of biometric technologies into the biopolitical configuration of U.S. empire, war and bodies. The contours of this matrix of empire, war and bodies are marked by what Woodward (2005, 32) calls an "approach that is multitheater, multiservice, multifunctional, and multibiometric." In his mapping of the complex dimensions of the biopolitical state, Foucault (2008, 77) observes that "The state is nothing else but the mobile effect of a regime of multiple governmentalities." The U.S. administration's emphasis on the importance of the "multi-"—in terms of multiple technologies that produce, in their interoperability, mobile apparatuses of biopolitical governance and securitisation—graphically exemplifies Foucault's point. Placed in the service of the biopolitics of identity management, remote biometric technologies (such as facial-scan and gait recognition) are being mobilised in order to monitor both watchlists and blacklists of target subjects (see Li, Schouten and Tistarelli 2009, 8–10).

In the multitheatres of U.S. imperial war, multibiometric systems aspire to differentiate "friend from foe." Gait signature, iris-scan, facial-scan, finger-scan—all are now mobilised in this multiservice exercise of imperial power. The term "frontier technologies" positions these new technologies within the historical context of expansionary relations of colonial and imperial power: *frontier* technologies signify both the inexorable teleological progress of Western technology and the consequent deployment of new technologies in securing and consolidating the borders or frontiers of empire. In the midst of multitheatre imperial U.S. wars, and their deployment of multibiometric systems, are the occupied bodies of colonised subjects. In this context, these colonised bodies are captured, biometrically scanned and their data is networked within U.S. interoperational databases. The U.S. Department of Defense (DoD) has established the Automated Biometric Identification System (ABIS) for both military and counter-terrorism purposes. Under the rubric of "Science of Security: How Do You Distinguish Friend from Foe," the DoD articulates the aims of ABIS:

> With the DoD ABIS, biometric data (with an initial focus on fingerprints) gathered by our forces from "red force" personnel—detainees, internees, enemy prisoners of war and foreign persons of interest as national security threats—can be compared with data maintained by the FBI's integrated automated fingerprint identification system (IAFIS), an electronic, searchable database with the fingerprints of approximately 48 million people who have been arrested in the US. Databases of US government agencies will also eventually be linked so that red force biometric data is searched against multiple databases for any possible matches.
>
> (Woodward 2004, 112)

A recent issue of the *New York Times* published a photo of U.S. troops taking a retina scan of an Iraqi citizen with the gloss: "Trying to Distinguish Friend From Foe On Baghdad's Streets: An American soldier took a retina scan yesterday to identify an Iraqi after the search of his house in Mansour, a western Baghdad neighborhood that has been torn apart by sectarian violence" (2007, A7). Violence in Baghdad is here scripted as something indigenous to the Iraqi people indulging their native predisposition to "sectarian violence" (for a critique of the terms "insurgents" and "sectarian violence" in relation to the war in Iraq, see Osuri 2009). Nothing is mentioned here of the imperial war unfolding in Iraq and the catalogue of massacres of Iraqi citizens (Jamail 2007), or of the years of U.S.-imposed sanctions that crippled the possibility for Iraqi citizens to mobilise against Saddam Hussein's regime; nothing is mentioned here of the more than 600,000 civilian Iraqi dead that have been documented as victims of the U.S.-led Iraqi war (Tavernise 2006, 11).

Situated within the elided context of imperial war, the *New York Times* photograph graphically represents the exercise of invasive power and biometric control by the U.S. military over the bodies of Iraqi citizens. This image of U.S. troops forcing an Iraqi citizen to undergo a retina scan evidences the material instantiation of imperial *geocorpographies* of violence. The term *geocorpographies* marks the indissociable relation between geopolitics, bodies and biopolitical technologies of inscription, surveillance and control (Pugliese 2007). Iraqi citizens no longer have autonomy or control over their own bodies, more specifically over the biometric identifiers of their own bodies. In this theatre of war, the bodies of Iraqi citizens become geocorpographically coextensive with the imperially invaded and controlled terrain of Iraq. In the "colonized sector," writes Fanon (2004, 4–5), "you die anywhere, from anything. It's a world of no space ... The colonized sector is a sector that crouches and covers, a sector on its knees, a sector that is prostrate. It's a sector of niggers, a sector of towelheads." The violence of colonial occupation evacuates space: there is no space; rather, in this colonised sector, bodies become coextensive with space as such: they are the ground upon which military operations are performed and through which control of the colonised country is secured. Coextensive with the imperially occupied terrain of Iraq, these subjugated bodies are compelled to submit to a systemic exercise of biopolitical power that mines their biometric signatures. The corporeal network of blood vessels that identify a subject's retinal signature becomes the physiological terrain that is biometrically mapped, identified, classified and colonised. Biometric retina scan, in this context, operates in terms of a "political intervention-technique with specific power effects" (Foucault 2003, 252). As a biopolitical technology, iris scan effectively disciplines the body of the subjugated Iraqi civilian through the enforced prising open of her or his eyelid; simultaneously, the scanned template is inserted within a networked grid of biopolitical intelligibility which claims to identify "friend from foe" and thereby sorts population groups according to an imperially imposed series of categories and classifications ("friend," "foe," "insurgent," and so on) determined by the "moral epistemology of imperialism" (Said 1992, 15).

The invasiveness that is represented in the *New York Times* photograph (of Iraqi citizens compelled to undergo retina scans) ramifies on a number of levels. In the photograph, the eye of the Iraqi citizen is literally prised open by a U.S. soldier, who keeps the eyelids separated as another secures a biometric scan of his retina. As subjects living under the regime of imperial occupation, Iraqi citizens are non-citizen subjects within their own nation. The physical prising open of a U.S. citizen's eye in order to obtain a retina scan would constitute immediate grounds for a lawsuit. In contradistinction, Iraqi bodies must be made available to the demands of the occupying power. This is the "open access" to both terrain and bodies that imperial powers demand and exact from their occupied subjects. Virtually existent across multiple U.S. databases, these *in silico* bio-signatures are digitally mobilised in order to secure identity dominance. On the ground, in the harrowed context of war, the lived reality of these occupied bodies of Iraqi citizens constitute another form of biometric signature altogether—a painfully lived signature inscribed by violence and shadowed by death.

Biometric Templates as Proxies for Criminalisation

In his analysis of surveillance post-September 11, David Lyon coins the critical term "digital discrimination." Digitial discrimination "consists of the ways in which the flows of personal data—abstracted information—are sifted and channelled in the process of risk assessment, to privilege some and disadvantage others, to accept some as legitimately present and reject others," and this is "increasingly done *in advance* of any offence" (Lyon 2003, 81). As I discuss in some detail later, in the context of Project Hostile Intent, the crux of the problem in the discriminatory deployment of biometric technologies post-9/11 resides in the manner in which discrimination occurs "in advance of any offence." The process of being criminalised in advance of having committed any offence compels the targeted subject to inhabit the fraught modality of the *quasi-prior*. The modality of the quasi-prior enunciates the manner in which targeted racialised subjects (such as those identified by the ethnic descriptor "of Middle Eastern appearance") are, already in advance of their actual identities, erroneously pre-comprehended and often unjustly apprehended by technologies and figures of surveillance (Pugliese 2003; Parenti 2003, 204–5). As Lyon (2003, 133) explains, in the context of a post-9/11 climate, "systems [are] being developed ... to create a 'threat index' for each passenger."

In biometric systems, processes of identification and authentication of enrolled subjects are predicated on generating a template or digital proxy of the subject. Encoded in this logic, however, is a series of contradictory, if not altogether aporetic, effects that problematise liberal-humanist conceptualisations of both identity and the subject. The aporetic logic of citationality (that I discuss in detail in Chapter 4), whereby the veridictional status of a subject's re-enrolling template is adjudicated precisely by its failure exactly to coincide with the original enrolment template, inscribes univocal conceptualisations of identity and the subject with a heteronomous law of the self-same-as-other. Indeed, the very status of the key signifiers of biometric identification and verification—uniqueness and authenticity—are predicated on an unacknowledged dependence on the other: the self-same subject must generate a micrological series of citations-as-differentiations that de-totalise her or his identity, even as these citations-as-differentiations function to affirm the seeming univocality of identity. Operative here, in effect, is an aporetic logic within which a "true" and "authentic" identity must be simultaneously, at the time of re-enrolment, non-identical to itself—in other words, at some minimal level, the re-enrolling scan must appear as a "fraud" in relation to the original enrolment template.

The western legal category of the subject is founded on the Enlightenment conceptualisation of identity as univocally fixed and self-same. As Stuart Hall (1993, 275) argues, the Enlightenment subject is founded on the notion of a centred and unified individual "whose 'centre' consist[s] of an inner core which first emerged when the subject was born, and unfold[s] with it, while remaining essentially the same—continuous or 'identical' with itself—throughout the individual's existence. The essential centre of the self was a person's identity." Biometric systems are predicated on this Enlightenment understanding of the subject: the authorising logic of these systems is driven by the notion that, despite micro permutations, the empiricity of flesh (the iris, the face or the fingerprint) encodes an identity that is continuous or identical with itself throughout the subject's existence. So, for example, the retina is celebrated in biometric literature because its "intricate network of blood vessels is a physiological characteristic that remains stable throughout the life of a person" (Nanavati et al. 2002, 108).

This understanding of identity as fixed and self-identical is clearly evidenced in the U.S. Department of Defense's conceptualisation of biometric identity and its strategic military import:

> The Department of Defense, with the full support of the White House, has recognized the collection of biometric identification as a basic warfighting capability, especially when fighting insurgent enemies who hide among the civilian populations ... So when a terrorist is captured in the field or a safehouse is raided, it is important to 'freeze' the terrorist's identity so that he can always be identified as an enemy and a potential threat.
>
> (Office of the Under Secretary of Defense for Acquisition, Technology, and Logistics 2007, 39)

This notion of "freezing" a subject's identity is founded on the Enlightenment view of identity as constituted by a self-identical and unchanging biological core. Yet within biometric systems this logic must be underpinned, simultaneously, by dissemination and decentring of self-same identity. This other, heteronomous logic is perfectly encapsulated in the following biometric formula: "Identification is often referred to as *1:N (one-to-N* or *one-to-many*), because a person's biometric information is compared against multiple (N) records" (Nanavati et al. 2002, 12). One-to-N or one-to-many names the manner in which the unitary identity of the subject is already invested with the law of the other (signatory citation as different in every instance). It is this heteronomical law that guarantees the conditions of possibility for the self-same to be constituted as an identifiably unique identity, even as it opens up the same to a movement of discontinuity and dissemination (across various institutional sites and biometric systems with every instance of re-enrolment).

I want to pursue this movement of discontinuity and dissemination of identity in the context of biometric technologies by proceeding to theorise biometric enrolment templates in terms of scattered "body-bits"—and I deploy the term "bits" in both its *in silico*, computer discourse sense and in its metaphorical meaning of segmenting and anatomising the body of the biometrically scanned subject. Biometric enrolment templates are not, effectively, body-bits at all: they contain no biological material; rather, they algorithmically transmute corporeal features into digital data. However, this digital data is nothing if not a tropological version of a subject's body-bits (iris, retina, fingerprints and so on): no fleshly body, no biometric template; decorporealised body-bits are predicated on the corporeality of the subject's "distinctive physiological characteristics" (Nanavati et al. 2002, 10). Biometric technologies tropologically transpose corporeal attributes to digital data. The very movement of this transposition from one to the other produces the rhetorical turn of tropology: biometric templates are tropic proxies of the body, specifically, as I discussed in Chapter 2, they are synecdoches of the subject—body-bits that signify the whole subject.

This tropic turn must be seen as instantiating what Emmanuel Levinas (1998, 162) terms "a denucleation of the very atomicity of the one." This process of denucleation of the atomicity of the one-as-self-same has far reaching consequences in its effective solicitation of the Enlightenment conceptualisation of the legal category of subject. Not only does it disrupt the unique identity of the self-same subject by disclosing an "irreducible category of difference *at the heart* of the Same" (Levinas 1998, 162), it also opens up a fundamental scission between the subject and the exercise of agency over their very body and their identity. I want to examine the ramifications of this scission in an analysis of the biometrics of racial profiling and the global search for terrorists.

As outlined in an article titled "Irises, voices give away terrorists," the U.S. government is utilising the Biometrics Automated Toolset (BAT) in its hunt for terrorists. BAT deploys biometric scanners to process a person's features, including their irises and voices; a series of algorithms is then used in order to convert the features to digital data. Put in Levinasian terms, the biometrical reduction of the face of the other to digital data is tantamount to the imperial project of "annexation by essence": that is, the other's body is fragmented into so many body-bits (for example, "fingerprints gathered from, say, drinking glasses or magazine covers found in terrorist haunts" [Associated Press 2002]) before being reconstituted into the digitised image of an essentialised unicity that ensures that "vision moves in to grasp" (Levinas 1988, 191).

Within this economy of dispersed body-bits, the relationship between the subject and their exercise of agency over their body is fissured. For example, against either the intentionality or will of the subject, a subject's body-bit, once converted to a biometric template, will effectively disclose the identity of the clandestine or dissimulated subject. Hostage to the timbre of its voice or the pattern of its irises, the body offers itself up despite the subject.

In this theatre of war, the fragmented body is compelled, unilaterally, to denounce and surrender the subject. In biopolitical regimes, the goal of absolute knowledge/power always returns to the empiricity of the flesh. In the folds of the body's skin, in the tonality of its voice, in the physiology of its eye, is incarnated an identity that cannot be dissimulated—even as the bits of flesh are already disassembled and converted to *in silico* matter that no longer resembles the subject that it proceeds to identify. The identity of the subject comes into being in this very movement of dissimulation: already in its digital conversion, as an array of binary numbers, it is non-identical to itself. The contemporary digital technologies that are instrumental in enabling this cleavage between body and subject are the culmination of that complex array of precursor technologies genealogically traced in Chapter 1. Discussing the panoply of nineteenth-century instruments developed "objectively" to evidence the criminal through "unmediated" recording of the brain or the pulse, David Horn (2003, 86) writes:

> If the body of the scientist (marked by its sensory weakness and the imprecision of its utterances) receded into the background, the body of the experimental subject was at the same time foregrounded and given a new agency. Whether the brain "wrote" or the pulse was "armed with a pen," the body was imagined to tell its own truth, and to give itself away.

I draw attention to military, defence and surveillance uses of biometrics in order to begin to map the contradictory forces at work in the space designated by the postfoundational subject: even as the subject has been deconstructed, decentred and dispersed, an opposing biopolitical labour is assiduously working to ground and centre the subject in the context of the most reductive biologico-positivism. Within these political economies of biopower, race continues to play a fundamental role as an instrumentality and technology of "identity dominance." In the contemporary biopolitical context, race has been atomised only to be reconstituted, for example, at the bio-molecular level of genes and "ethnic specific alleles" that code for a subject's "racial" identity (Pugliese 1999). The biopolitical deployment of genetics in the imperial quest for "identity dominance" is evidenced by recent moves by the U.S. Department of Defense to use the Armed Forces DNA Identification Laboratory and the Armed Forces DNA Repository (originally established to manage the "mortuary affairs of the Department of Defense" in terms of forensic identification of military personnel) in order to identify prospective terrorists. Code-named operation "Black Helix," this DNA repository "will focus on archiving, retrieving, and interpreting biomolecular data for the identification and tracking of terrorist suspects" (Office of the Under Secretary of Defense for Acquisition, Technology, and Logistics 2007, 31).

Biometrics interweaves flesh with algorithms in order to archive so many body-bits that reconstitute identity regardless of a subject's permutations. This biometrical conceptualisation of identity falls squarely, as I outlined above, within Enlightenment theories of identity and the subject. Produced by the conjoining of the genes of one's parents at the moment of conception, a subject's genetic identity is conceptualised as essentially remaining the same throughout the individual's existence: "The DNA molecule at the heart of each cell in the human body is like a signature, unique to each individual" (Aldridge 1996. 187). Inscribed in the inner core, in the "heart of each cell," a subject's genetic identity is homogenous and self-identical to itself. Yet precisely because it is inscribed, analogically and thus tropically, as "signature," as textual sequence (for example, AGAATTC), its biological status cannot but be discursively and tropologically constituted (Pugliese 1999). At the very moment when, within the discourse of genetics, a DNA profile is identified and named as unique, as proper to a particular subject, it is always already divided in and of itself: the differential marks that constitute the conditions of possibility for the emergence of the subject's unique DNA signature simultaneously interrupt its self-identity. This is not to deny the empirical status of a DNA profile; rather, it is to underline the manner in which a DNA identity, as a scientific positivity, is the result of discursive and textual processes generative of material effects.

Within the schema of the Biometric Automated Toolset, the language of algorithms operates effectively as an ontological "reduction of the other to the same by the interposition of a middle and neutral term that ensures the comprehension of being" (Levinas 1988, 43). The interposition of the scientific term—for example, "biometrics" or "Screening Passengers by Observation Technique"—is supposed to guarantee the objective and neutral dimensions of this process, while effacing the ongoing regimes of racial profiling that continue to inflect these same projects. For instance, the U.S. is now deploying Transportation Security Administration (TSA) officers across its major airports. The surveillance role of these "behavior-detection" officers, through the use of a program called SPOT (Screening Passengers by Observation Technique), is "to discern the subtlest suspicious behaviors" in order to screen and capture prospective terrorists: "The TSA considers the program a powerful tool to root out terrorists, but also an antidote to racial profiling" (Shukovsky 2007). Yet, as Naseem Tuffah, political chairman of the American Arab Anti-Discrimination Committee's Seattle chapter, has warned: "if the TSA 'only looked hard when somebody is Middle-Eastern-appearing ... then you are still conducting racial profiling under a different name'" (quoted in Shukovsky 2007).

The critical transformations at work in the biometric dissemination of identificatory templates and the consequent scission between body, subject and identity that I have so far outlined need to be further elaborated. In his analysis of the liberal-humanist subject, Levinas (1988, 36) tracks the movement of what he terms identificatory "self-recovery" in the face of the vicissitudes of change and transformation: "The I is not a being that always remains the same, but it is the being whose existing consists in identifying itself, in recovering its identity throughout all that happens to it … The I is identical in its very alterations. It represents them to itself and thinks them. The universal identity in which the heterogeneous can be embraced has the ossature of a subject, of the first person." This movement of self-recovery of one's identity falters, however, at those points where the ossature of a subject is in fact fractured and denucleated by the processes of biometric replication and networked-database classification and dissemination. The self-recovery of one's identity no longer remains the unilateral operation of the "first person."

Biometric templates exemplify the post-biological, *in silico* networking of a subject who can no longer govern or control the heterogeneous dispersal of her or his identificatory body-bits-as-template-proxies. It is at this juncture that a disarticulation is enunciated, a disarticulation that vitiates a subject's agentic self-recovery and governance of her or his heterogeneous body-bits and identity proxies. At this juncture, a subject's biometric proxies may be mobilised, indeed, as agents of the biopolitical state deployed to ensnare and convict the targeted individual. "We are there," writes Levinas (2003, 67), "and there is nothing more to be done, or anything to add to this fact that we have been entirely delivered up"—delivered up by our biometric proxies to the state. Malcolm Crompton (2002) has examined the privacy implications of the biometric capture of bodily information, with a particular focus on the manner in which a person's biometric scans can often reveal a range of medical conditions. Within the specific context of racial profiling, as deployed by both military and police forces, I want to emphasise that a subject's biometric body-bits may become precisely *proxies for criminalisation.* "When being black (or Latino or Asian [or "of Middle Eastern appearance"] is used as a proxy for criminality or dangerousness in a society in which relative few are criminals," David Harris (2002, 106) notes, "profiles based on or including race will always sweep too widely."

Within these political economies of biopower, biometrics interweaves flesh with algorithms in order to "freeze" a subject's identity regardless of his or her permutations. At work here are the biopolitical operations whereby, to paraphrase Levinas, the subject is riveted to the fatality of the biological, even as the biological is transmuted *in silico*: "The essence of humanity is no longer in freedom but in a kind of bondage. To be truly oneself is … to become aware of the ineluctable bondage unique to your body … And then, if race did not exist, it would be necessary to invent it!" (Levinas cited in Rolland 2003, 31). The re-invention of race in the context of this biometric riveting of identity to the body is clearly enunciated in these contemporary modalities of racial profiling and, as I discuss later, in the biometric surveillance and control of geopolitical borders, refugees and asylum seekers.

Biometrics at the Border: Converging Biometric Technologies and the Surveillance and Control of Asylum Seekers and Refugees

The war on terror has generated the mobilisation, by western governments, of a wide range of scopic technologies across multiple sites, both civilian and military, in the construction of surveillance, identification and screening networks. What is unfolding, in effect, is the convergence of technologies of surveillance with biometric systems of identification and authentication in order to constitute a new regime of "counter-terrorism" enabling hyper forms of governmentality, surveillance and biopolitical control. For example, the Council of Australian Governments, after convening a "Special Meeting on Counter-Terrorism," proposed a national approach to closed-circuit television together with an investigation of "the means by which reliable, consistent and nationally interoperable biometric security measures could be adopted by all jurisdictions" (Council of Australian Governments 2005, 2, 5).

The move towards scopic technologies of interoperability emerges as another form of biopower. As discussed in the Introduction, biopower draws on technologies designed to bring the body into the field of calculations for the purposes of governance, surveillance and control. Within this field of calculations, the individual body becomes the pivotal point of application for regimes of biopower. Under the rubric of "Targeting," the Report of the Defense Science Board Task Force on Defense Biometrics states that: "Our military and intelligence concerns in the Global War on Terrorism have largely shifted away from the nation states and their facilities, and towards individuals" (Office of the Under Secretary of Defense for Acquisition, Technology, and Logistics 2007, 2). The U.S. Department of Defense's preoccupation with "identity management" clearly articulates its biopolitical agenda in marking a shift of focus away from nation states to embodied individuals or, to use Foucault's biopolitical term, "somatic singularities." The atomising of the body of the individual through processes of biometric screening works to secure "identity dominance."

The anatomy of biopolitical power is, in the contemporary context, constituted literally and metaphorically by governmental uses of biometric technologies. On the one hand, the literality of the body—the individuating singularity of its biometric signature—is what enables the very operating logic of biometric technologies. On the other hand, the dispersed and multiple *in silico* bodies of subjects—digitised biometric templates—function to constitute a metaphorical anatomy of power, an anatomy of biopolitical power technologically networked in order to facilitate ever more rigorous regimes of governmentality and surveillance under the mantle of "identity management systems." In the words of the Report on Defense Biometrics: "An identity management system, here, is meant to include both algorithms, their instantiation in both software/hardware, as well as data. The data are an organized collection of information about specific individuals. Indeed, when we ask 'who are you,' we are really asking 'what are you—e.g., friend or felon?'" (2007, 8). As I discussed in Chapter 1, in the context of the imbricated development of biopolitics and biometrics, the defining biometric question—"Who are you?"—is fundamental to the operations of the biopolitical state. And the question "Who are you?" is clearly predicated on a politico-ontologised conceptualisation of identity in which *who you are biopolitically defines what you are*.

In the U.S. Report of the Defense Science Board Task Force on Defense Biometrics (Office of the Under Secretary of Defense for Acquisition, Technology, and Logistics 2007, 8), the management of identity, and the consequent production of identity dominance, is seen as something that can be achieved through the establishment of biometric identity databases:

> It is easiest to think of an identity database as a relational database, rows and columns, where the rows ("entities" or "records") are individuals, where the columns (or "attributes") are characteristics or categories of information about individuals, and where the column entries (or fields) represent the particulars for that individual. Certain attributes serve principally to "identify" you, that is, to allow one to query (or "index into") the database and retrieve some or all of your record. Among traditional "identifiers" are name and social security number (SSN). Names may be our first impulse, but they are notoriously ambiguous and generally not sufficiently unique. SSN is more unique. All of these variables, however, suffer from the problem that they can be compromised relatively easily—bought, stolen, or invented. Thus, they are increasingly insufficient, by themselves, for identification. That brings us to biometrics.

A biometric identity database is seen as offering a "truly accurate Identity Management system" precisely because it can cross-reference "rows" with "columns" of seemingly non-replicable bodily signatures (Office of the Under Secretary of Defense for Acquisition, Technology, and Logistics 2007, 15). Here the anatomy of biometric biopolitical power can be seen to be discursively enabled by a computational grid within which a subject's identificatory attributes (physiological, genetic, behavioural) are distributed along a paradigmatic axis (columns) and the targeted individual (the "entity" or "record") is aligned along a syntagmatic axis. This biometric computational grid is seen to offer the "root identity." The "root identity" of an individual, the Report (2007, 37) claims, is "the ground of truth":

> An important distinction here is the difference between "true identity," a unique, provable, fact for which the only real proofs are biometric in nature; and a "persona" that one may adopt as being appropriate to some kind of identity-sensitive activity, such as sending e-mail or conducting an online auction. The easy distinction is that an Identity is an

> irreducible core fact, while a Persona, if it [is] to be trusted, should have recourse to a true or "root ID," whether or not that is visible to all parties, all the time.
>
> (2007, 14 n 8)

As I discuss in detail in Chapter 4, a subject's biometric "root identity" is, in keeping with the rhetorical effects of this rhizomatic trope, always caught within a transversal movement of iterable repetition that precludes the possibility of an authoritative self-identity not always already marked by difference. A subject's biometric "root identity" is as susceptible to fraud and imposture as any other signatory identity. However, what is significant here is the manner in which the concept of "true or root ID" achieves its conditions of intelligibility by being conceptually grounded on regimes of veridiction and biologised/essentialised notions of identity. The transposition of these biopolitical-biometric regimes of identity management to the border, in order to screen, regulate and exclude target populations, is what I now want to consider.

In tracing the development of fingerprint biometrics, Simon Cole (2002, 136) has underlined how the technology was quickly taken up by immigration authorities in the U.S.: "Fear of the foreigner and of the racial and ethnic 'other' remained, at the turn of the twentieth century, the primary application for fingerprint identification." Cole proceeds to examine the critical relation between fingerprint biometrics and the screening of racially profiled immigrant populations. From its inception, in other words, biometric technology has played an instrumental role in the biopolitical governance of suspect foreigners at the borders of the nation. In the U.S., the federal government enacted, between 1996 and 2002, a series of laws underpinning the development of the Chimera system:

> an automated information system, to gather and share information among agencies about aliens seeking to enter or stay in the United States. The major requirements for this Chimera system are (1) biometric identification; (2) machine-readable visas, passports, and other travel documents; and (3) interoperability among all State Department, INS [Immigration and Naturalization Service], and federal law enforcement and intelligence agency systems that contain information about aliens. (U.S. Government Accounting Office 2002)

The U.S. has simultaneously deployed the IDENT computer system for the identification and surveillance of immigrants. With the IDENT system "digitalized 'biometric' photographs and fingerprints ... can be searched for matches among millions of other such images ... [T]he IDENT system is now used throughout the new Department of Homeland Security to identify and track immigrants both at the borders and inside the United States" (Parenti 2003, 178–79; see also van der Ploeg 2005, 123–24).

In Europe, member states of the European Union have, in the form of the "Eurodac information system," deployed biometric technologies at the border in order to screen prospective asylum seekers:

> The Eurodac system enables Member States to identify asylum seekers and persons who have crossed an external frontier of the Community in an irregular manner. By comparing fingerprints, Member States can determine whether an asylum seeker or a foreign national, found illegally present within a Member State, has previously claimed asylum in another Member State. In addition, by being able to check if an applicant has already lodged a request for asylum in another Member State, "asylum shopping" in other Member States after being rejected in one can be avoided.
> (EURODAC [European Dactyloscopie]Supervision Coordination Group 2007, 4)

In her analysis of the Eurodac system, van der Ploeg (1999, 299) has drawn attention to the manner in which undocumented asylum seekers, in being compelled to be enrolled within fingerprint databases before their claims for asylum can be processed, are *a priori* branded with the "mark of illegality" because of the criminalising status of fingerprint biometrics. In the Australian context, the federal government is currently testing the use of biometric systems for the purposes of border control through the identity management of refugees and asylum seekers. I quote from Amanda Vanstone's (former Minister for Immigration and Multicultural and Indigenous Affairs) media release:

> The trials involve a state of the art biometrics system ... Biometrics can take much of the guess work out of [the] ... checking process. The trial tests our capacity to record, store and match biometric information effectively and efficiently using two main groups of volunteers. The first group involves overseas travellers selected for further checking by immigration officials at the airport. Biometric information, including fingerprints, iris scans and facial information will be collected from volunteers. This will test our capacity to effectively and efficiently record and store information. (2005)

The use of the term "volunteers" strategically occludes the asymmetries of power that are operative in government officials "requesting" biometric information from travellers "selected for further checking." These asymmetries of power are marked even more graphically when considering the other group of "volunteers":

> The second group comprises refugees travelling to Australian from Africa. These volunteers have had their biometric information recorded

> before departure. When they arrive in Sydney they are asked to again provide their biometric details allowing us to test our capacity to match the information. The trial is the second step in developing our capacity to use biometrics ... The airport trial and others this year are the first live trials in operational areas. In the future, we could also take biometric information from people turned around at the border and add this information to our alert lists.
>
> (Vanstone 2005)

One wonders exactly what "choice" refugees—who, by definition, are desperate to flee their country of origin—have in declining to offer up their biometric information to Australian immigration officials in Africa. What needs to be underscored here is the fact that refugees, in these biometric trials, have been coupled with travellers who are under governmental suspicion: "overseas travellers selected for further checking." The insensitivity of using this most vulnerable and traumatised of all groups of travellers, refugees, to take part in governmental trials is only outdone by the manner in which they are, by implication, placed in the same trial category as "suspect" travellers. The structurality of this categorical inclusion is made clear in the context of the criminalisation-through-imprisonment of "unauthorised" refugees who arrive in Australia. This process of imprisonment upon arrival is driven, as Suvendrini Perera (2002) has argued, by the government's argument that "unauthorised" refugees could be "stand-ins for terrorists." This criminalisation of refugees by association is further evidenced by including them in the same discussion of future uses of biometric information taken "from people turned around at the border"—a phrase now permanently inscribed in the annals of Australian history with the scandal of *Tampa*, as well as all the other boats laden with refugees refused entry under the regime of the "Pacific Solution" (see Perera 2002, 2002a, 2006, 2008, 2009). As Dean Wilson argues (2007, 211), "The particular focus on asylum seekers in biometric deployment is intertwined with the securitization of migration and the criminalization of persons seeking refuge in Australia."

This process of criminalising refugees has been further elaborated by federal government legislation that now demands that all asylum seekers and their children be "included in a criminal database, with their fingerprints to be rolled into the national system" (Welch 2008). This will give the:

> National Automated Fingerprint Identification System (NAFIS) access to the Immigration Department's database of about 3000 detainee fingerprints ... Police investigating a crime can seek information, including name, address and immigration history, about a current or former detainee from the department if they score a "hit" on the database, which holds 4.2 million sets of fingerprints. It is managed by the

> national criminal information agency, CrimTrac. About 90 per cent of the fingerprints are from convicted criminals.
>
> (Welch 2008)

As Dr Mohammed Al Jabin, an Iraqi international law expert, has observed: "These people are not criminals. They are to be treated in accordance with the rules of the Refugee Convention and they are not to be listed as a criminal or even included on a list of criminals" (quoted in Welch 2008). Despite these human rights protestations, the Australian government has proceeded with this biometric plan.

On yet another front, the United Nations High Commission for Refugees (UNHCR) has, since October 2002, been using iris-scan biometrics "to verify the identities" of Afghan refugees "seeking assistance to return to Afghanistan" (UNHCR 2003). This seemingly neutral exercise in the use of biometrics on refugees is in fact underpinned by racialised presuppositions about Afghans that are imbricated with paternal relations of welfare. The refugee's iris-scan is checked against a database holding over 200,000 biometric templates of Afghan refugees. According to UNHCR (2003), "If the code has not appeared before, the refugee is registered and given clearance to receive an assistance package on arrival in Afghanistan ... If the test reveals that the refugee has been enrolled before—only about half of one per cent are found to be 'recyclers'—the person is refused assistance." Iris-scan, in the context of UNHCR's economy of paternal welfare, functions to expose the miserable scandal of utterly disenfranchised "recyclers" asking for "second helpings." Rae Lynn Schwartz-Dupre (2007, 444) succinctly names and identifies the biopolitics at work here:

> The criminal nature of Afghani refugees is the motive for the technology. UNHCR reports that "the system has detected approximately 1,000 people who have tried to claim assistance for a second time" ... And "Afghan refugees eyeing a second helping of UNHCR's reparation assistance have hit a blind spot with the arrival of state of the art iris recognition technology" ... Thus, people everywhere can rest assured knowing that refugees fleeing their homes will not get second helpings. The crime of "recycling," as the UNHCR refers to it, is one that falls in line with racist and classist beliefs that the poor are criminal by nature.

Evidenced in the deployment of iris-scan biometrics by the UNHCR in order to screen Afghan refugees is that crossing over of the biopolitical question ("Who are you?") with the biometric question ("Who is X?") that I drew attention to in Chapter 1, with the attendant conflation of who you are with what you are. As I argued earlier, the manner in which the veridictional question of biopolitics is embedded within a biometric conceptualisation of identity as innate (ontologised) is succinctly evidenced by this formulation of biometrics: "What you are: Biometrics" (Woodward et al. 2003, 7). In this specific context, the biometric of iris-scan confirms the biopolitical status of Afghan refugees: that the Afghan "poor are criminal by nature" and that they thus need to be digitally screened, identified and sorted. It is precisely the conflation of "what you are" with "who you are" that is celebrated by the U.S. General Accounting Office's report on the use of biometrics for border security: "The use of biometrics—things the travellers are—can more securely bind a person's identity to a travel document" (2002).

The racialogical theories of criminal anthropology and their imbrication with biometric technologies, as I examined in Chapter 1, are, in the contemporary context, continuing to play crucial biopolitical roles both at national borders and internally, specifically in the surveillance of "suspect" population groups within western nations. Indeed, the seemingly anachronistic belief in the criminality of an "entire stock" is exactly what has informed the recent passing of legislation by the Italian parliament ordering the fingerprinting of all Roma people living in Italy, including children. As Mary Gibson (2002, 157) writes, in her critical analysis of criminal anthropology's belief in the "born criminal," "By arguing that born criminality was visible on the body, criminal anthropology encouraged the idea that police could correctly identify deviants even before they broke the law." In contemporary Italy, police have been rounding up and fingerprinting *all* Roma women, men and children as a contemporary exercise in "preventive criminology," producing, in effect, an "ethnic register that would treat Roma children as if they were hardened criminals" (BBC 2008).

The convergence of technologies of surveillance, identification and authentication into networked systems of hyper-surveillance and biopolitical control must be seen as operating at two critical levels: the legislative and the technological. At the legislative level, for example, the passing of the Australian *Migration Legislation Amendment (Identification and Authentication) Act 2004*, that came into effect on 27 August 2004, provides

> a wider legislative basis for collecting personal identifiers such as photographs, signatures and fingerprints, to enhance DIMIA's [Department of Immigration and Multicultural and Indigenous Affairs] ability to establish and authenticate the identity of non-citizens at various stages of immigration processing, and on entry to and departure from Australia.
>
> (DIMIA 2004)

The development of networked systems of bipolitical control and hyper-surveillance is being implemented under the rubric of what DIMIA calls an "integrated approach":

> DIMIA is also working with the Department of Foreign Affairs and Trade and the Australian Customs Service to design an integrated approach to the use of biometric technology for border control.

> As part of this process, DIMIA is developing an Identity Services Repository for the matching of face and fingerprint images for visa applicants.
>
> (DIMIA 2004)

The Identity Services Repository extends the networked reach of governmental anatomies of power even as it operates as an anatomising archive of biometric-templates-as-body-bits.

At the technological level, this legislative convergence of technologies of surveillance, identification and authentication is being enhanced by another form of "integrated approach": multibiometric systems. These systems signal the possible overcoming of the limitations of unimodal biometric technologies that I discussed in Chapter 2. Rather than relying on capturing a single biometric trait (such as a face-scan), multibiometric systems rely on multimodal biometric capture (such as scans of face, fingers, irises, gait, palm print and so on). Consequently, "Multibiometric systems are expected to enhance recognition accuracy of a personal authentication system by reconciling the evidence presented by multiple sources of information" (Ross and Poh 2009, 288). In keeping with governmental moves towards an "integrated approach" in terms of networking technologies of surveillance, authentication and identification, "multibiometric systems can integrate information at various levels" and "are expected to be more reliable due to the presence of multiple, fairly independent pieces of evidence" (Jain and Ross 2004, 6, 3).

Biometric Pre-Emptive Surveillance: Project Hostile Intent and the Integrated Score of Malfeasance Likelihood

The mobilisation of biometric technologies in order to secure identity dominance can be seen to be moving into the realms of science fiction through the funding and development, by the U.S. Department of Homeland Security, of Project Hostile Intent (PHI). PHI will be based on "video cameras, laserlight, infra-red, audio recordings and eye-tracking technology [that] are expected to scour crowds looking for unusual behaviour, with the aim of identifying people who should be approached and quizzed by security staff" (Sample 2007). Designed as a multimedia and multimodal system, PHI is being designed to "pick up tell-tale signs of hostile intent or deception from people's heart rates, perspiration and tiny shifts in facial expressions" (Sample 2007). Coextensive with the science fiction realm represented by the Hollywood film *Minority Report*, and its technologies of pre-emptive surveillance (Lyon 2008, 149), the goal of PHI is to "identify people 'involved in possible malicious or deceitful acts'—before they ever commit the crime" (Eisenberg 2007). The Department of Homeland Security envisions integrating PHI with biometric technologies and their networked databases.

With PHI, the suspect subject is enmeshed within a regime of biopower designed to measure, calculate and gauge the body's most minute physiological and behavioural manifestations. Pulse rate, perspiration and micro facial expressions are all digitally calibrated and gauged against what one of the computer scientists involved in the project has called an "integrated score of malfeasance likelihood" (NewsCenter 2007). Presupposed in this integrated score of malfeasance likelihood is a disciplinary norm that establishes the guiding templates for "correct," "normal" and "appropriate" behaviour against which deviations signal targets that pose security risks. Thus subjects suffering from mental illness, those under medication, those who, due to fraught and traumatic histories of racial profiling and police harassment, reactively break into a sweat at the sight of police or other law enforcement officials, are all positioned as suspect subjects who breach the integrated score of normative behaviour and its regulatory mechanisms. PHI perfectly reproduces the functional attributes of biopower and its technologies of "infinitesimal surveillances" (for example, micro-facial expressions) and its mechanisms of "micro-power concerned with the body" (Foucault 1990, 145–46).

PHI also fulfils the operational requirements of the apparatuses of biopower: "to qualify, measure, appraise, and hierarchize" the subjects under its multimodal system of surveillance, effecting "distributions around the norm" (Foucault 1990, 144). PHI is designed to *qualify* behaviour and physiological signs (increased perspiration or pulse rate is due to anxiety generated by having one's criminal intentions exposed); it both *measures* and *appraises* these same corporeal signs and effects, and then proceeds to *hierarchise* them along a disciplinary axis of normative behaviour. As Foucault (1990, 144) argues, the law, within these regimes of biopower,

> operates more and more as a norm, and ... the judicial institution is increasingly incorporated into a continuum of apparatuses (medical, administrative, and so on) whose functions for the most part are regulatory. A normalizing society is the historical outcome of a technology of power centred on life.

Governing the operational logic of PHI is the conceptual *a priori* of the norm. In PHI the norm is not a fixed and singular quality; rather, it functions as "an interplay of differential normalities" (Foucault 2007, 63), in which a series of corporeal indices are gauged against a dynamic range of infrastructural normalities. Moreover, PHI's measuring of an individual's non-normative micro-facial expression or pulse rate works to constitute the target subject in terms of a quasi-criminal whose criminal intent has been exposed by a series of "infralegal analogies": micro-facial expression or pulse rate operate as "kinds of miniature warning signs ... that are presented as already analogous to the crime" (Foucault 2003a, 23). In other words, PHI's dependence on regimes of tacit, quasi-illegality functions to constitute incipient criminal subjects (for example, figures "of Middle Eastern appearance") who already resemble the crimes they have not actually committed but are in danger of committing in some indeterminate future. With PHI, the subject in question is placed in an infra-legal space that situates her or him as suspect before any actual infraction of law. As such, this infra-legal space enables law to extend its operations into the amorphous domain of discretionary power premised on the ability to decode a series of corporeal signs and preincident indices that will give away the prospective criminal before the fact. This amorphous domain of discretionary power and infra-legality is precisely what enables the practice, for example, of racial profiling to be conducted against target subjects: a series of visible indices, that is, racially coded phenotypical features, alert the police officer to a prospective criminal in advance of the fact of the subject having perpetrated any offence.

PHI, in claiming to extract a subject's "real" thoughts or intentions despite her- or himself, must be seen as performing a type of inquisitorial extortion of the "truth." This extortion of truth is effectively one of the key modalities of the biopolitical state. As discussed earlier, the biopolitical state is marked by the installing of the veridictional question, "the question of truth," at the core of its criminological operations. PHI is a biopolitical technology of extortion because it scans, evaluates and judges the target subject's physiological effects as signs of intent regardless of the subject's will. As an extortionary biopower apparatus, its reliance on the authority of science and expert medico-legal knowledge generates normative truths that discipline and punish those who fall outside the circumscribed parameters of the "integrated score of malfeasance likelihood." Furthermore, in the context of these emergent technologies, the relationship between the subject and their exercise of agency and control over their body is fissured: the scanned and processed body-bit—heart rate, perspiration and micro shifts in facial expression—is compelled, unilaterally, to denounce and surrender the subject.

PHI takes this process of splitting the body from subject one step further, as designated corporeal signs are employed in terms of empirical attestations of *intent*, that is, of acts that have not actually taken place but that might. In other words, target subjects are positioned corporeally to disclose and physiologically confess criminal intent in advance of having perpetrated a crime. PHI surveils both the surface of the skin (facial expressions) and the body's interior physiological processes (pulse rate) in order to make manifest what would otherwise remain invisible: intent. What is operative here, in effect, is the techno-visualisation of the intangible and invisible: mental thoughts directed at prospective or potential acts. It is at such a juncture that biopower can be seen clearly to be imbricated with disciplinary power. Disciplinary power is characterised by the need to keep subjects within states of "perpetual visibility" so that it can be exercised immediately the mere "intent" of a criminal act is suspected/detected. "Disciplinary power,"

Foucault (2006, 51–52) notes, "always tends to intervene beforehand, before the act itself if possible ... at the level of what is potential, disposition, will, at the level of the soul." Potential, disposition, will, intent, soul—these are the amorphous ectoplasms that PHI can claim to detect. The last in this series, the soul, brings into focus the confessional apparatus that, spectrally and genealogically, underpins PHI, animating its disciplinary-moral drive to extort the truth of the subject in question.

In splitting the body from the subject, PHI positions an individual's body as *obligated* to confess the true intentions of the subject. In regimes of biopower, Foucault (1990, 60) explains,

> The obligation to confess is now relayed through so many different points, is so deeply ingrained in us, that we no longer perceive it as the effect of power that constrains us; on the contrary, it seems to us that truth, lodged in our most secret nature, "demands" only to surface.

Lodged in the inner sanctum of the mind/brain, the intent of the subject scanned and evaluated by PHI is extorted against any act of agentic voluntarism through the somatechnical production of a docile body obligated to confess. All of this "unfolds," Foucault (1990, 61) writes,

> within a power relationship, for one does not confess without the presence (or virtual presence) of a partner who is not simply the interlocutor but the authority who requires the confession, prescribes and appreciates it, and intervenes in order to judge [and] punish.

With PHI, the scanning technologies become the virtual interlocutors who, in the first instance, extort the corporeal signs of malfeasant intent. Once exposed as a criminal suspect with malfeasance likelihood to commit a crime, the target subject is delivered over into the hands of the authority that will proceed to judge and punish the prospective criminal.

The biopolitical logic of a technology such as PHI is predicated on the future anterior: in the future, as evidenced by the captured scan of a subject's pulse or perspiration rate and evaluated against a normative behaviour template/integrated score, a criminal act will be judged to be that which has *mentally* already taken place. The arresting image of invisible intent, made involuntarily visible through designated corporeal signs, extorts the "truth" from a docile body despite the will of the subject.

4 Identity Fraud And Imposture: Biometrics, The Metaphysics Of Presence And The Alleged Liveness Of The "Live" Evidentiary Body

Biometrics, Power and Tactics of Resistance

So far in the course of this book, I have theorised biometric technologies as fundamentally constituted by relations of biopower, arguing, in the process, that the key biopolitical question, "Who are you?", animates the entire operating logic of biometric systems; and that, furthermore, this question repeatedly folds into its biopolitical counterpart, so that "Who you are" becomes coextensive with "What you are." As technologies of biopower, I have traced the heterogeneous uses to which these technologies have been put by governments, NGOs, corporations and so on in the constitution and governance of specific subjects (refugees, terrorists, suspect populations). Having framed biometric technologies as key instrumentalities in the exercise of the biopolitics of subjugation, surveillance and control, in this chapter I want to shift the focus somewhat. Specifically, I want to pose another Foucauldian question when faced by regimes of power: What tactics and resistances can be deployed so as to elude or short-circuit these regimes of power? I situate this question within Foucault's (1980, 142) theorisation of power as at once subjugating and productive of in-built resistances:

> hence one should not assume a massive and primal condition of domination, a binary structure with its "dominators" on one side and "dominated" on the other ... [T]here are no relations of power without resistances; the latter are all the more real and effective because they are formed right at the point where relations of power are exercised.

Taking this theorisation of power as my point of departure, I proceed to examine a series of techniques that might be deployed by fraudsters in order to trick biometric systems into giving them illegitimate physical and/or symbolic access to data and/or controlled areas. My aim is not to endorse the exercise of biometric fraud through techniques of spoofing; rather, it is to evidence the Foucauldian thesis of power as marked by in-built resistances and productive of unintended effects.

Biometric systems are being increasingly deployed across a wide range of institutions and organisations in order to provide security of access. Biometric systems are used to provide "*logical access* to data or information"; "*physical access* to tangible materials or controlled areas"; as well as identifying or verifying "the identity of an individual from a database or token" (Nanavati et al. 2002, 144). Across much of the relevant literature, biometric technologies are presented as providing virtually foolproof systems of identification and/or verification of the subjects that are biometrically screened. Yet, despite these claims, the literature also acknowledges that biometric systems are open to being tricked by fraudsters. In order to counter the tactics used by fraudsters to fool biometric systems, biometric scientists and technologists are in-building within the technologies a number of tests designed to detect fraudsters. One of the key fraud detection methods being deployed by biometric systems is so-called "liveness testing." Liveness testing is being used to determine whether the person being screened by the system is actually present (and "alive") rather than a simulacrum reproducing a stolen identity. In the course of this chapter, I proceed to situate the fraud detection methods of biometric "liveness testing" within a Derridean critique of the metaphysics of presence in order to disclose the philosophical presuppositions that inform scientific and technological understandings of the body and identity.

Deconstructing the Metaphysics of Presence

As discussed in Chapter 2, once a biometric technology, such as facial scan, has been set up in a particular context, the subject whose identity will be verified by this biometric system is required initially to enrol by supplying the requisite biometric data, such as a digital scan of their face, which is subsequently converted into a template. The template, that is generated by the algorithmic encoding of a subject's distinctive biometric features, is stored in the system and is used to verify a user's identity every time they present themselves for biometric screening—in other words, the initial enrolment template is matched against the user's verification template.

In what would seem, in the first instance, to be a counter-intuitive logic, biometric "enrolment and verification templates should never be identical": "Because different templates are generated each time a user interacts with a biometric system, there is no 100 per cent correlation between enrolment and verification templates" (Nanavati et al. 2002, 19, 21). Indeed, an exact one-to-one identical match is seen as the sign of fraud, as it signals the possibility that an impostor has stolen the initial enrolment template of someone else and is presenting it in order to illegitimately gain access to the system. This seeming counter-intuitive logic, that demands iteration of identity with difference, graphically exemplifies the deconstructive movement of iteration, as a movement always already inscribed with alterity in every new instance of repetition. Jacques Derrida discusses this paradox precisely in the context of that exemplar of unique identity: the *signature*—a term that is now a fundamental signifier in the discourse of biometrics, where it is used to name the unique identity of enrolled subjects across diverse biometric systems, including gait-signature, keystroke-signature and so on.

Derrida (1990, 20) unpacks the paradox of the signature in the context of deconstructing its representation as a privileged signifier that marks an indissociable tie to an originary figure, what he terms the "tethering to the source":

> In order for the tethering to the source to occur, what must be retained is the absolute singularity of a signature-event and a signature form: the pure reproducibility of a pure event ... But the conditions of possibility of those effects is simultaneously, once again, the condition of their impossibility, of the impossibility of their rigorous purity. In order to function, that is, to be readable, a signature must have a repeatable, iterable, imitable form.

The constitutively repeatable/iterable status of identity is, in fact, inscribed in the very etymological emergence of the term "identity." In his detailed tracking of the history of identity papers and passports in early modern Europe, Groebner (2007, 26) writes:

> "Identity" is a medieval coinage. It was in common use in its Latin form *idemptitas* or *identitas* in medieval logic. Derived from *idem*, "the same," or *identidem*, "time and again," it denoted not uniqueness, but the features that the various elements of a group had in common."

This embedded etymological meaning, Groebner (2007, 219) demonstrates, "[f]rom the mid-fifteenth century onwards," becomes the animating logic of technologies of identity verification and authentication: "It was reproduction that literally created the proofs of a person's individuality: an individual had to be doubled by an identity document plus an official internal record on the document issued."

In biometrics, this iterable and repeatable identity form can never be identical across each instance of its repetition: "As opposed to an identical string of data," Nanavati et al. (2002, 262) explain,

> biometric templates vary with each finger placement, iris acquisition, and voice recording: the same finger, placed over and over again, generates a different template with each placement. This is attributable to minute variations in presentation—pressure, distance ... which lead to the extraction of slightly different features for each template.

In other words, the unique identity biometric of a subject is indissociably tied to iterability, as "the logic that ties repetition to alterity" (Derrida 1990, 7). It is the logic of iterability that problematises biometrics' reliance on the foundational concept of the "root identity," as "The authoritative identity [of a subject] established and maintained with high integrity by the system" (Office of the Under Secretary of Defense for Acquisition, Technology, and Logistics 2007, 163). In the biometric literature, a subject's "root identity" is described as being predicated on the "ground truth" (Parziale and Chen 2009, 107) of their biometric attributes. A subject's biometric "root identity" is, however, in keeping with the rhetorical effects of this rhizomatic trope, always caught within a transversal movement of iterability that precludes the possibility of an authoritative self-identity not always already marked by difference.

Conceptualised in Derridean terms, then, the biometric signature of a subject can only function as the instantiation of a unique signature-event that signals the "presence" of a non-fraudulent subject through an iterable movement that must be differentially marked at every "presentation"—as the "process by which a user provides biometric data to an acquisition device" (Nanavati et al. 2002, 17). I place "presence" in scare marks as there is inscribed in this aporetic movement—in which the identity of the self-same must be at once different—not only a deconstruction of the concept of a unique identity indissociably tethered to a root source, but there is also inscribed a deconstruction of the metaphysics of presence that fundamentally informs the discourse of biometrics and its constitutive lexemes.

Before I proceed further in my discussion of the constitutive role of the metaphysics of presence in biometric technologies, I want to spend some time unpacking Derrida's deconstruction of the metaphysics of presence. The identity of a subject or sign is constituted, in Derridean terms, through the operations of *différance*, that is, the identity of the sign "cat," for instance, only achieves its signifying value through a system of differential relations where the letters "c," "a," and "t" differ from all the other letters of the alphabet: "every concept is inscribed in a chain or in a system within which it refers to the other, to other concepts, by means of the systematic play of difference" (Derrida 1986, 11). This differential relation, Derrida argues, is also constituted by a complex system of deferrals, in that all the other signs of the alphabet that differ from the letters "c," "a," and "t" are at once deferred in their "appearance" even as they are constitutive, through their difference, of the signifying value of "c," "a," and "t":

> It is because of *différance* that the movement of signification is possible only if each so-called "present" element, each element appearing on the scene of presence, is related to something other than itself, thereby keeping within itself the mark of the past element, and already letting itself be vitiated by the mark of its relation to the future element, this trace being related no less to what is called the future than to what is

> called the past, and constituting what is called the present by means of this very relation to what it is not ... An interval must separate the present from what it is not in order for the present to be itself, but this interval that constitutes it as present must, by the same token, divide the present in and of itself, thereby also dividing, along with the present, everything that is thought on the basis of the present, that is, in our metaphysical language, every being, and singularly substance or the subject.
>
> (Derrida 1986, 13)

Derrida (1986, 11) here draws attention to the manner in which the present can never be fully present unto itself as it is always already divided by this play of difference and deferral: "The first consequence to be drawn from this is that the signified concept is never present in and of itself, in a sufficient presence that would refer only to itself." Such foundational categories of western metaphysics as "being" or "the subject," then, can no longer be thought as constituted by a self-identical presence; rather, these foundational categories, premised on a metaphysics of pure and undivided presence, must be seen as the effects of this play of *différance*. This detour into Derridean deconstruction has been essential in order to begin to disclose the manner in which biometric systems are underpinned precisely by this unacknowledged metaphysics of presence; and, furthermore, the ever-present danger of spoofs and frauds is actually, I argue, a system-effect of biometrics' reliance on a metaphysics of presence. The one (a metaphysics of presence) produces the other (frauds, impostors).

Biometric Latency and Bogus Authentication

In Woodward et al.'s *Biometrics: Identity Assurance in the Information Age* (2003, 8), a fictional character named "Cathy" is constructed by the authors in order to illustrate how bogus identification and identity spoofing can occur within biometric systems:

> The biometric authentication process begins with a biometric sensor of some kind. When Cathy tries to log in, the sensor collects a biometric reading from her and generates a biometric template from the reading, which becomes the authenticator. The verifier is based on one or more biometric readings previously collected from Cathy. The verification procedure essentially measures how closely the authenticator matches the verifier. If the system decides that the match is "close enough," the system authenticates Cathy; otherwise authentication is denied.

Woodward et al. (2003, 8) call the measured properties of Cathy's biometric trait "the base secret in a biometric system." This "base secret," however, turns out to be, through another aporetic turn, publicly available as a type of "latency":

> it's important to recognize that her [Cathy's] biometric traits aren't really secrets. Cathy often leaves measurable traces of these "secrets" wherever she goes, such as fingerprints on surfaces, the recorded sound of her voice, or even video records of her face and body. This "latency" provides a way for attackers to generate a bogus authenticator and use it to trick the system into thinking that Cathy is actually present. Moreover, it may be possible to intercept a genuine authenticator collected from Cathy and replay it later. Thus, accurate authentication depends in part on whether the system can ensure that biometric authenticators are actually presented by live people.
>
> (Woodward et al. 2003, 8)

The public latency of the base secret encapsulates the aporetic logic of biometrics as a *somatechnology*. As I discussed in the Introduction, somatechnics refers to the indissociable relation between bodies (*soma*) and technologies (*technè*): bodies can only achieve their cultural intelligibility, precisely as "bodies," through their inscription by various technologies, including language. The animating principle of all biometric systems is the technologisation of the body's key identifiers: the body's identificatory features must be extracted and technologised into digital templates. Somatic features, within biometric systems, are only intelligible once they have been algorithmically processed and "fixed" into templates. This process enables the serialisation of biometric features as the logic of the system is predicated, as I argued earlier, on the iterability of unique identificatory features or, couched in Derridean terms, on the possibility of "originary reproduction" (Derrida 1976, 209), in which the unique becomes culturally intelligible as "unique" through the effaced process of its very iterability. And this dependency on the iterable logic of originary reproduction is not something exclusive to biometric technologies of identity. Rather, it is constitutive of all identification-based systems. As Groebner (2007, 252) concludes in his survey of the history of identity documents in early modern Europe, "the history of identification is at once the history of the technologies of reproduction." Although he does not draw on Derridean theory to explicate his argument, Groebner effectively articulates a deconstructive understanding of the philosophical presuppositions that underpin identity and reproduction. Remarking critically on what he terms "the fiction of authenticity," Groebner (2007, 219) underscores how the earliest identity-based documents were, unsurprisingly, already haunted by the spectres of impostors, fraud, and proxies:

> The history of identification I have traced from the mid-fifteenth to the end of the seventeenth century leads to an unequivocal conclusion.

> After two centuries of regulation, laws, and ever newer forms of official documents declared compulsory, after two centuries of bureaucratic orders—"Register everyone and everything!"—and of repeated admonitions that stricter attention be paid to recording and checking individuals, what was the outcome of all these endeavours? The rise of the con man and the impostor … Their careers in dissimulation took place not in spite of, but through the expanding systems of bureaucratic control.

It is, then, the structural demand that unique identificatory features be reproducible/iterable, in order to be biometrically intelligible and legible, that generates the very possibility for fraud. The measurable traces of the biometric "secrets" that a subject leaves behind function to construct a public "theatre" of latent spoofs, spectres and feints. In the course of the practices of her or his everyday life, a subject leaves a trail of biometric traces (fingerprints, DNA, CCTV images) across diverse spaces and contexts. These traces are at one and the same time secrets that are publicly available to be put to use in a repertoire of feints and impostures. The aporia that I am marking here of a secret that is simultaneously public is in fact constitutive of the logic of the secret as such. This is the "enigma" of the secret that Derrida (1992, 95) draws attention to:

> The enigma of which I am speaking here … is the *sharing of the secret*. Not only the sharing of the secret with the other, my partner in a sect or in a secret society, my accomplice, my witness, my ally. I refer first of all to the secret shared *within itself*, its partition "proper," which divides the essence of a secret that cannot even appear to one alone except in starting to be lost, to divulge itself, hence to dissimulate itself, as secret, in showing itself: dissimulating its dissimulation.

In order for a secret to be a secret as such it must institute a "negation that denies itself" (Derrida 1992, 95): dividing itself ("its partition 'proper'"), the secret must "lose" itself, divulge itself as already other, as "dissimulating its dissimulation" in order to maintain its status as secret. The secret's status as secret is predicated on its denegated disclosure and dissimulation of itself. Graphically inscribed here, then, is the aporetic logic that animates the possibility for all the biometric secret traces that a subject leaves behind to be dissimulated by another. One's "proper" somatic traces are—as latently legible biometric traits that are, in Woodward et al.'s (2003) words, "read" by biometric systems—only legible because they are simultaneously iterable and inscribed by alterity, that is, by the mark of the other.

I draw on the metaphor of a public "theatre" of biometric spoofs and mimics as the techniques available to subvert biometric systems are all couched in a performative lexicon of masquerade and mimicry. Woodward et al. list the following techniques of biometric fraud:

> *Masquerade*: This is the classic risk to an authentication system. If Henry's [another fictional character] goal is masquerade, he's simply trying to convince the system that he is in fact someone else, perhaps Cathy, since the system already knows how to recognize her. Henry proceeds by trying to trick the system into accepting him as being the other person.
>
> (2003, 9)
>
> *Replication*: In this attack, Henry produces a copy of whatever Cathy is using to authenticate herself.
>
> (2003, 13)
>
> *Mimics*: Mimics are when a user is able to impersonate another identity. (2003, 14)
>
> *Artifacts*: Artifacts are when an attacker is able to present a manufactured biometric (such as a fake finger) to the system.
>
> (2003, 14)
>
> *Digital Spoofing*: Also known as a *playback attack*, this attack takes advantage of the fact that all authentication data is ultimately reduced to bits on a wire. If the system expects a particular value for the authenticator, the attacker intercepts this value and replays it to masquerade as someone else.
>
> (2003, 14)

Operative in this theatre of biometric mimics and impostors is what Derrida (2002b, 57) terms, in another context, a "mediatic-techno-performativity and a logic of the phantasmata." The "authentic" and "unique" biometric signature of a subject can only be rendered legible biometrically by being mediated by the operations of digital technology. The process of biometric authentication can only be staged through a performative of mediated iterability that is at all times open to the haunting spectres of latent phantasmata: unique "bits" of the subject as so much discarded but latent traces waiting to be capitalised, "re-animated," by the fraudster-in-waiting. The logic of iterability that underpins all biometric systems establishes the conditions of possibility for both the technological encoding of the unique features of a subject's *soma* and the reproducibility of these unique traits as so much mediated techno-digital data: "The possibility of repeating," writes Derrida (1990, 8), "and thus of identifying the marks implicit in every code, making it into a network [*une grille*] that is communicable, transmittable, decipherable, iterable for a third, and hence for every possible user in general."

The key signifiers that underpin all biometric technologies and that are constitutive of their system of conceptuality are *trace*, *secret* and *iterability*. I would argue, at this juncture, that the aporetic logics of iterability, the trace and the secret vitiate any claims that a foolproof system can ever be built that is not also always already open to frauds and impostors. The very possibility of fraud and imposture is in-built within the unacknowledged metaphysics of presence of biometric systems. In what follows, I want to elaborate on this unacknowledged metaphysics of presence by focusing on so-called "liveness" testing in order to preclude instances of biometric identity fraud.

Signs of Life: The Alleged "Live" of Liveness Testing

In their chapter titled "Biometric Liveness Testings," Valorie S. Valencia and Christopher Horn (2003, 10) bring into focus the unsettling spectre of spoofs that haunts biometric systems:

> Recent reports have shown that biometric devices can be spoofed using a variety of methods ... The security provided by biometric devices—that is, the level of confidence in the user's identity—is diminished if the devices can be readily circumvented. Liveness detection, among other methods, has been suggested as a means to counter these types of attacks.
>
> *Biometric liveness tests* are automated tests to determine if the biometric sample presented to a biometric system came from a live human being—not just any live human being, however, but the live human being who was originally enrolled in the system—the "authentic live human being," if you will.
>
> One way to defeat a biometric system is to substitute an artificial or simulated biometric sample for the biometric sample of the "authorized live human being." As such, liveness testing is a technology used to maximize confidence that individuals are who they claim to be, and that they are alive and able to make the claim.

"The fundamental faith of the metaphysicians," Nietzsche notes (1966, 10), "is *the faith in opposite values.*" Inscribed in the aforementioned Valencia and Horn passage is a metaphysics founded on the faith in opposite values: authentic/fake, live/dead, present/absent. These opposite values are presented as foundational categories that can be empirically verified. This metaphysical faith in opposite values is precisely what is undone by biometrics' dependency on the logic of iterability. Animating biometrics' liveness tests is a metaphysics of presence, in which an "authentic live human being" presents herself before the technology. Undivided from herself, fully in possession of her "proper" and "authentic" traits, the subject undergoing the liveness test presents herself in the full plenitude of her self-identical "liveness." These biometric fraud detection tests are underpinned by the metaphysics of presence and, precisely because it is a metaphysics, it fails to deliver what it promises. As I argued earlier, for the biometric traits of a subject to be rendered legible as a biometric template within the system, they must assume the form of an iterable mark or signature. As iterable mark, a subject's biometric signature is always already inscribed by *différance*, in which the self-same is at once deferred and different from itself (Derrida 1986, 8–9). A subject's biometric signature must conform to a grammatological understanding of "writing" that presupposes the "death" of a subject even as they present themselves "live" before the biometric system; in other words, the subject must "go through the detour of the sign" (the enrolment template lodged in the biometric system) in order to be intelligible as identifiable subject by the biometric system:

> To be what it is, all writing must, therefore, be capable of functioning in the radical absence of every empirically determined receiver in general. And this absence is not a continuous modification of presence, it is a rupture in presence, the "death" or the possibility of the "death" of the receiver inscribed in the structure of the mark ...
>
> What holds for the receiver holds also, for the same reasons, for the sender or producer. To write is to produce a mark that will constitute a sort of machine which is productive in turn, and which my future disappearance will not, in principle, hinder in its functioning, offering things and itself to be read and to be rewritten.
>
> (Derrida 1990, 8)

As I demonstrated earlier, the authenticating and identificatory logic of biometric systems is predicated on generating a template proxy of the subject. Encoded in this process is a series of aporetic effects that problematise liberal-humanist conceptualisations of both identity and the subject. The aporetic logic of iterability and citationality, whereby the veridicity of a subject's re-enrolling template is adjudicated precisely by its failure exactly to coincide with the original enrolment template, inscribes univocal conceptualisations of identity and the subject with a heteronomous law of the self-same-as-other. Indeed, the very status of the key signifiers of biometric identification and verification—uniqueness, authenticity and veridicity—are predicated on an unacknowledged dependence on the other: the self-same subject must generate a micrological series of citations-as-differentiations that de-totalise her identity, even as these citations-as-differentiations function to affirm the seeming univocality of identity.

As I discussed in Chapter 3, biometric systems of identification and verification are predicated on an Enlightenment understanding of the subject that conceptualises the self as stable and continuous throughout the course of a person's life. Yet, this concept of a self-identical self is problematised by a constitutive process of dissemination and decentring of self-same identity. The identity of the subject comes into being in this very movement of dissimulation: already in its algorithmic conversion, as an array of digital numbers it is non-identical to itself. This is the crux of the matter underpinning the constitutive effects of the metaphysics of presence in the operations of biometrics: the somatic identity of the subject cannot present itself in the purity of an "uncontaminated" *physis*, that is, in terms of a body not always already culturally marked and discursively inscribed as a legible *body*. An uncontaminated and irreducibly pure *physis* would remain non-iterable and thus "illegible." In talking, then, of the somatechnics of biometrics, I am drawing attention to the iterable and thus tropic (because proxy-prosthetic) status of a subject's biometric identity/signature. As discussed in my Introduction, in articulating the aporetic hinge between the originary and the reproducible (Pugliese 2005a, 362), between the natural (*physis*) and the synthetic-prosthetic (*technè*), Derrida (2002a, 244) emphasises that the relation "is not an opposition; from the very first there is instrumentalization [*dés l'origine il y a de l'instrumentalisation*] ... a prosthetic strategy of repetition inhabits the very moment of life. Not only, then, is technics not in opposition to life, it also haunts it from the very beginning." If a prosthetic strategy of repetition-as-instrumentalisation inhabits the very moment of life, then liveness testing can never fully circumvent the deployment of faux body parts and synthetic-prosthetics: the possibility of the proxy already marks and constitutes the very possibility of the body proper. Indeed, no body proper as such that is not always already inscribed by instrumentalisation and its prosthetic inscription by cultural systems and techniques of signification.

In their article, "Body Check: Biometric Access Protection Devices and Their Programs Put to the Test," Lisa Thalheim, Jan Krissler and Peter-Michael Ziegler(2002) have put to the test eleven biometric systems by generating digital spoofs and proxies and, in all cases, they managed to breach the system's security screening devices. In their tests, these researchers have outfoxed biometric protective programs and devices by "deceiving the systems with the aid of obvious procedures (such as the reactivation of latent images) and obvious feature forgeries (photographs, videos, silicon fingerprints)" (Thalheim et al. 2002). As they document in their article, they obtained "astonishing results by means of this approach" (Thalheim et al. 2002). Thalheim et al. proceeded successfully to spoof the biometric systems they tested by supplying a range of relevant biometric simulations. For example, in order to breach the liveness test of a facial scan system, they:

> simply shot a short.avi video clip with the webcam in which a registered user was seen to move his head slightly to left and right. As brief movements suffice for FaceVACS to consider an object alive and as the program engages in simple 3D calculations only, we were not particularly surprised by the success of our approach: Once the appropriate display-to-ToUcam distance had been found the program did in fact detect in the video sequence played to it a moving "genuine" head with a known facial metric, whereupon it granted access to the system.
>
> In a worst case scenario this state of affairs implies that a person without a professional background to movie making who had wielded a digital camera during a public meeting and there shot visual material of

> authorized personnel, to log on to a protected system, need only modify the acquired material slightly and transfer it to a portable PC.
>
> (Thalheim et al. 2002)

The fact that Thalheim et al. are compelled to place the term "genuine" (head) in scare marks highlights the aporetic logic that haunts and inscribes biometric systems predicated on the binary oppositions "genuine" and "fake," "live" and "dead," "body" and "machine." And I re-iterate the following Derridean citation in order to elaborate my critique of the metaphysics of presence in the context of biometric systems: "What holds for the receiver holds also, for the same reasons, for the sender or producer. To write is to produce a mark that will constitute a sort of machine which is productive in turn, and which my future disappearance will not, in principle, hinder in its functioning, offering things and itself to be read and to be rewritten" (Derrida 1990, 8). As the production of an identificatory mark/signature is dependent upon the "disappearance" of the signatory subject (in order for the signature to be able to function as proxy in-lieu of the absent subject), this structural disappearance or "death" of the signatory subject is what ensures the very production of the biometric proxy *and* the possibility to trick the system with a dissimulation of the simulation. The structural need to produce a proxy of the subject at biometric enrolment generates the "possibility of disengagement and citational graft which belongs to the structure of every mark ... Every mark can be cited, put between quotation marks" (Derrida 1990, 12). In the context of biometric systems, the act of identity theft hinges precisely on the logic of the citational graft: the thief purloins another subject's biometric signature and presents it to the system with the hope that it will fail to read the quotation marks.

"An advantage of biometric authentication technologies," write Valencia and Horn (2003, 142) in their "Biometric Liveness Testing," "is that we can do something about it—we can incorporate automated liveness tests to minimize the effectiveness of artificial or simulated biometric specimens." The sort of entanglement of contradictory terms that is evidenced in this "solution" pervades the discourse of biometrics: "automated liveness tests" signals, paradoxically, the automated machining of the live, the technologisation of the body in order to attempt to differentiate between *soma* and *technè*, the living flesh and the dead simulation or prosthesis. Yet this metaphysics of pure and unmediated presence is incessantly undone by the fact that the liveness of the "here-now does not appear as such, in experience, except by differing from itself" (Derrida 2002b, xvii). The "traits" that Derrida (1990, 9, 10) grammatologically

> recognize[s] in the classical, narrowly defined concept of writing, are generalizable. They are valid not only for all orders of 'signs' and for the entire field of what philosophy would call experience, even the

> experience of being: the above-mentioned 'presence' ... there is no experience consisting of *pure* presence but only chains of differential marks.

The experience of pure, undifferentiated and non-technologised being is what underpins biometrics' system of conceptuality. The biometric system deploys an "automated liveness" test in order to detect "*signs* of life": in other words, the *in vivo* must be rendered *semiotically in silico* in order to register as a "sign of life." The "bio" of bioinformatics is only ever available, as a culturally intelligible unit of information, through the *indissociable* transposition or transcoding of the one (bio) into the other (informatics)—that is, through an ineluctable process of somatechnicity. The impossibility of the experience of unmediated being is underscored by the fact that the logic of general citationality constitutes the biometric system's unacknowledged conditions of possibility. The possibility of digital spoofing, as a form of "structural parasitism" (Derrida 1990, 17), is structurally in-built in the system. And, in placing biometric security systems through their paces, Thalheim et al. (2002) deploy a range of both serious and farcical tactics of parasitism and citational grafting, including breathing over the trace of a fingerprint in order to "revivify" it, grafting a fingerprint trace onto adhesive film, reactivating a latent image with a water-filled plastic bag or balloon, and deploying an inkjet print of a human eye perforated with a miniature hole in order to trick an iris-scan system.

In their discussion of biometric templates, Kari and Landweber (2000, 414) argue that there is an "homology" between a subject's biometric template/signature and his or her DNA signature: "The complex structure of a living organism ultimately derives from applying a set of simple operations (copying, splicing, inserting, deleting, and so on) to initial information encoded in a DNA sequence." This homology, in fact, resonates along a number of levels. Biometric traits are viewed in terms of a subject's unique genetic and/or phenotypical features: "biometrics rely on genetics as the basis of various biometrics" (Woodward et al. 2003, 29). As I have argued elsewhere, in my grammatological deconstruction of genetics, DNA is only intelligible as a scientific object of inquiry through the deployment of a series of effaced metaphors predicated on writing, including genetic letters, codes, texts, polymerase proofreaders, spelling errors, traces and so on (Pugliese 1999). This homology between genetics and biometrics holds, not only because both disciplines are critically dependent upon a textual economy of writing and *différance* in order to make legible their respective objects of inquiry, but also because both disciplines are foundationally dependent on a metaphysics of "life" informed by an empirico-positivist biologism. In the fields of science and technology, whenever the problematic of "life" is invoked, a metaphysics of pure and unmediated biological presence is unreflexively called into "being." As Derrida (1976, 70) observes, "in all scientific fields, notably in biology, this notion [of presence] seems currently to be dominant and irreducible." Yet, as I demonstrated earlier, at the very moment that life, the living organism, is encoded in biometric language (or genetic text), it becomes inscribed in the movement of *différantial* deferral of and difference from the other; at the moment of biometric presentation "there is no experience of *pure* presence but only chains of differential marks" (Derrida 1990, 10).

"Biometric authentication," explain Woodward et al. (2001, 11), "refers to *automated methods of identifying or verifying* the identity of a *living person in real time* based on a *physical characteristic* or *personal trait*." The identificatory machining or automation of the living person in real time encapsulates the non-negotiable aporias that inscribe biometrics, in which a subject's biometric *image file* is termed the *corpus*—that is, in which the *image* of a subject's bio-identificatory feature becomes, indissociably, her machined/automated corpus. Biometrics' techno-automated mediation of the "live" and "real time" signifies, in effect, that there can only ever be "an allegation of 'live'" and of "real time." Discussing the metaphysics of presence in the contemporary configuration of "tele-techno-mediatic modernity" and its celebration of such things as "live" satellite-televisual transmissions, Derrida (in Derrida and Stiegler 2002, 40) sardonically remarks "we should never forget that this 'live' is not an absolute live, but only a live effect [*un effect de direct*], an allegation of 'live'." The allegation of live captures the metaphysics of presence that unreflexively inform biometrics' faith in liveness testing (as the means whereby to circumvent digital spoofs and identity frauds). It is an *allegation of live* that animates biometric liveness testing as "to allege" signifies, in legal terms, "to assert without proof" and, simultaneously, to "cite, quote" (*Shorter Oxford English Dictionary*). In other words, the very act of "live" authentication before a biometric system can only ever remain an assertion without absolute proof as it can only be performed through the grammatological process of template citation and signatory quotation, a process that irreducibly dissimulates both "life" and "authentic" identity and that structurally ensures that the "absolutely real present is already a memory": "there is no purely real time because temporalization itself is structured by a play of retention or of protention and, consequently, of traces ... The real time effect is itself a particular effect of 'différance'" (Derrida in Derrida and Stiegler 2002, 129).

This is not to reduce the liveliness of life to the operations of an homogenising textuality or totalising techno-discursivity. Rather, the liveliness of life must be viewed as what exceeds the algorithmic delimitations and empirico-positivist frames of the biometric sciences. "The *liveliness* of life," writes Levinas (1998, 162, 178), "is an incessant bursting of identification"; "Is not the liveliness of life an excession, the rupture of the container by the uncontainable?" The excession of the liveliness of life signifies the impossibility of containing a subject's life within the calculable parameters of digitised "identity" categories/templates: already non-identical to itself, the liveliness of life inscribes itself in so many citational grafts, structural parasitisms and heteronomous traces, thereby dissimulating itself and exceeding the metaphysics of presence.

Transductions of the Body, Infrastructural Normativities and Biometrics' "Extrinsic" Information

Because of its unacknowledged dependence on a metaphysics of presence, biometrics is structurally haunted by the threat of "spoofability." The spectre of this threat is what is driving the development of new biometric technologies designed to outfox frauds and impostors. A recently developed biometric system, vascular pattern recognition or vein pattern identification, is being touted as yet another system that "is difficult to forge": "vascular patterns are difficult to recreate because they are inside the hand and, for some approaches, blood needs to flow to register the image" (NSTC 2006, 1). Another emerging biometric technology that is being promoted as offering spoof-detection capabilities is finger skin histology, which entails the imaging, through the use of optical coherence tomography, of the "internal structure of the skin of the finger":

> The skin on the palmar side of the finger tips contains dermatoglyphic patterns comprising the ridges and valleys commonly measured for fingerprint-based biometrics. Importantly, these patterns do not exist solely on the surface of the skin—many anatomical structures *below the surface* of the skin mimic the surface patterns. For example, the interface between the epidermal and dermal layers of skin is an undulating layer made of multiple protrusions of the dermis into the epidermis known as dermal papillae. These papillae follow the same shape of the surface dermatoglyphic patterns and thus represent an internal fingerprint in the same form as the external pattern.
>
> (Nixon, Aimale and Rowe 2008, 414; italics original)

Biometrics' search to create a non-spoofable identificatory system has taken the technology below the surface features of the body (and its physiognomic and behavioural attributes) and into the seemingly non-replicable depths of the *soma*. Lodged in the depths of the *soma*, beyond the realm of replicable corporeal surfaces, is what Foucault (1975, 94) terms the "visible invisible," as that master metaphor that appears to promise a corporeal "truth" that is homogeneously self-identical and non-replicable. Vascular pattern recognition operates by "Using near-infrared light" so that "reflected or transmitted images of blood vessels of a hand or finger [or face] are derived and used for personal recognition" (NSTC 2006, 1). Yet, a subject's vascular patterns cannot simply signify in the self-evidence of their own unique corporeality. They must be "transduced," to use the apposite biometric term, by "combining geometry with underlying physiology" (Office of the Under Secretary of Defense for Acquisition, Technology, and Logistics 2007, 52).

The transduction of the physiological into the geometric is what enables the "visibilisation" of the invisible: the tropics of transduction, as fundamentally driven by the *turn* of metaphor, enable the somatechnic transmutation of the organic material of the body into intelligible data (that can be "read" by the biometric system) through a series of instrumental mediations. These mediations include the use of algorithms to remove "noise" (such as hair or shadows) and in order to "enhance" "the clarity of vascular patterns in captured images" (Choi and Tran 2008, 264).

There is, however, a prior process of mediation that antedates the mediations I have just identified. Critical to this process of rendering the organic articulate and intelligible is language, as that other technology that has already inscribed the body even before the process of biometric scanning and algorithmic transduction has begun. Vascular pattern recognition and finger skin histology technologies come to a body that has already been techno-linguistically mediated before the fact of infrared scanning and optical coherence tomography. Infrared and tomographic scanning of the body's "invisible visible" can only take place as a coherent and intelligible techno-scientific operation after the fact of the discursive medico-anatomical mediation of the *soma*. Before the process of biometric infrared scanning and optical coherence tomography, the internal physiology of a subject's body has already been mapped (in anatomical atlases) and identified and rendered intelligible through a series of medico-anatomical terms: vascular pattern, blood vessels, blood flow, subcutaneous blood vessel pattern, the haemoglobin of the blood, capillary tufts, dermal papillae, and so on. This medico-anatomical lexicon constitutes the conditions of possibility for biometrics to embark on its infrared and tomographic "descent" into the body's interior in order to extract its unique identificatory features. And, once again, these tropical transductions of the "raw" organic material of the *soma* cannot escape either the logic of iterability or its consequent spoofable effects.

As I discussed in Chapter 2, the somatechnics of life ensure that the contours of the subject's body do not terminate at the threshold of the body; rather the contours extend beyond the physical parameters of the subject into the larger sociocultural domain, inflecting the operating parameters of biopolitical technologies determining critical question of knowledge/power. At the threshold of the *soma*, that empirical point of seeming terminus dividing the corporeal from the non-corporeal, the flesh is metaphorised into its other: *technè*. Inversely, *technè* is metaphorised as *soma*: in other words, what takes place is a *somatechnics of biometrics*. Even as biometrics dreams of the development of ever-new technologies designed to differentiate between "authentic" and "fraudulent" subjects, the technology's entire system of conceptuality is haunted by the ineluctable spectre of its absolute unthought: that the *soma* has always already been technologised before the fact of biometric scanning and template creation.

This absolute unthought is brought into critical focus by the division between "primary biometric information" and "extrinsic" or "ancillary biometric information" that underpins biometrics' system of conceptuality. Primary biometric information refers to the face, fingerprint or iris that has been biometrically scanned and processed, whereas extrinsic biometric information refers to "characteristics such as gender, ethnicity, height or weight of the user (collectively known as soft biometric traits)" (Nandakumar, Ross and Jain 2008, 335). Posited as "extrinsic," "ancillary" and "soft," the inscriptive categories of ethnicity and gender are positioned as structurally separable from the biometrically scanned body. As such, these categories are what can be imported from "outside" of the scanned body as "add on" or "ancillary" information to the "primary" data of somatic identifiers: iris, face or fingerprint. Yet, as I demonstrated in the context of the non-random failure to enrol of particular ethnic subjects in the face of infrastructural whiteness, there is no such thing as a body that is not always already marked by a constellation of social descriptors (including ethnicity and gender) prior to the moment of biometric processing. These descriptors are not extrinsic to the body; on the contrary, they constitute the body's *a priori* conditions of social signification and cultural intelligibility and, consequently, position the subject in determinate ways in the face of particular biometric technologies, producing material effects: for example, failure to enrol.

There is, furthermore, in this lacuna or systemic unthought that inscribes biometric systems of conceptuality, a type of contradiction that results from this positing of categories such as gender and ethnicity as "extrinsic" to the biometric body in question. In their essay, "Incorporating Ancillary Information," Nandakumar, Ross and Jain (2008, 348) proceed to argue that: "Soft biometric traits are available and can be extracted in a number of practical biometric applications. For example, attributes like gender, ethnicity, age and eye color can be extracted with sufficient reliability from the face images. Gender, speech accent, and perceptual age of the speaker can be inferred from the speech signal." Framed as "extrinsic" to the biometrically scanned body, gender and ethnicity are simultaneously "extracted" from both the face and the voice of the subject as self-evident categories "lodged" in the body in question. Positioned as "ancillary" to the body, yet these categories self-evidently inscribe and identify their subjects. Thus, in the illustrations that accompany Nandakumar et al.'s (2008, 347) text, a photograph represents "A scenario where the primary biometric identifier (face) and the soft biometric attributes (gender, ethnicity, eye color and height) are automatically extracted and utilized to verify a person's identity." The photographed subject's gender is named as "male" and the ethnicity as "Asian." Underpinning the "automatic extraction" of such categories as gender and ethnicity are tacit and normative knowledges that will enable the "automatic" identification of these same categories. The self-evident or received status of these tacit knowledges is what enables the process of "automatic" identification.

In this biometric system of conceptuality, it is self-evident what a male or female "looks" like; it is self-evident what a male or female "sounds" like. In both these cases, the category of the gender variant subject, who might "sound" like a male but "look" like a female must remain unthought, as this subject falls outside the gender-normative assumptions encoded in the biometric system. And I invoke the gender variant subject not as some sort of eccentric anomaly that is marginal to the automated operations of biometric gender identification; on the contrary, at this critical juncture of automated, computational biometric gender identification, the gender variant subject, in crossing the self-evident attributes of heteronormative gender identities, effectively works to expose the occluded assumptions and tacit knowledges that proceed to inform the normative infrastructure/software of the biometric system. In her foundational work on transgender, Susan Stryker (2006, 3) underscores the power of transgender subjects to "reveal the operations of systems and institutions that simultaneously produce various possibilities of viable personhood, [whilst] eliminating others." In queering the heteronormative gender binaries that underpin the biometric system's process of automated gender identification, the gender-nonconforming subject is biometrically positioned as a subject that "does not compute," precisely as s/he brings into crisis the disciplinary operations of a system predicated on infrastructural heteronormativity and categorical, automated gender binaries.

The deconstructive force of the gender variant subject that I have invoked in this scenario is complicated, moreover, by the lived effects generated by the ongoing reproduction of normative gender categories, and the consequent gender misrepresentations and discriminations, across the broad spectrum of technologies and institutions assigned with identificatory tasks (Stryker 1994; Prosser 1998). These lived effects are documented, for example, by Toby Beauchamp in the context of the intensification, post-9/11, of surveillance of transgender bodies by the U.S. government. Beauchamp (2009, 359) draws attention to the manner in which gender-nonconforming bodies have been caught in the dragnet of the U.S. Social Security Administration's "'no-match' letters to employers in cases where their employee's hiring paperwork contradicts employee information on file with SSA." As Beauchamp (2009, 359) explains:

> The no-match policy aims to locate undocumented immigrants (and potential terrorists) employed under false identities, yet casts a much broader net. Because conflicting legal regulations often prevent trans people from obtaining consistent gender markers across all of their identity documents, gender-nonconforming individuals are disproportionately affected by the policy, whether they are undocumented immigrants or not.

Beauchamp's use of the "no-match" category can be productively transposed to a critique of biometric automated gender recognition systems, precisely where an effective "no-match" occurs between the normative gender assumptions of the operating software and the embodied, gender variant subject screened by the technology.

As with the question of gender, in this biometric schema of "extrinsic" or "ancillary" biometric traits the criteriological parameters that configure specific ethnic groups remain self-evident. There is no question at all that one can "automatically" identify someone who is "Asian," simply, I assume, by relying on stock taxonomies of visible phenotypical descriptors; in other words, a tacit ontology and taxonomy of visible racial attributes self-evidently signify one's ethnicity. And it is precisely at these junctures that such "extrinsic" identificatory information as gender and ethnicity are shown to be fundamentally inscribed as *a priori*, infrastructural normativities "intrinsic" to the classificatory and identificatory operations of biometric technologies. Even as biometricians labour to "import" these "extrinsic" attributes from outside the primary operations of biometrics, they are simultaneously shown to be always already inscribed on the body as self-evident, normative attributes that can be "automatically" extracted.

I end this chapter with one more moment of iteration: that the *soma* has always already been technologised before the fact of biometric scanning and template creation. It is the very technologisation of the *soma* that renders a subject biometrically legible as such. Consequently, in its critical reliance on the figure of the signature, a biometric system is permanently open to the possibility of citational grafts, structural parasitism and identity frauds. Biometrics is, as a techno-science, thoroughly dependent on a metaphysics of presence that is predicated on the onto-theology of unmediated essence; this is perfectly encapsulated in the biometric formula: "something you are, a biometric" (Woodward et al. 2001, 11). Finally, as biometrics' invocation of liveness testing remains nothing more than another animation of the metaphysics of presence, a mere allegation of "live," it can never absolutely guarantee that the figure before it is not life itself dissimulating its dissimulation.

5 Neurotechnologies Of Truth: Brain Fingerprinting's Neurognomics And No Lie MRI's Digital Phrenology

Neurotechnologies of Demonstrative Truth

The veridictional question, the question of "truth," is, as I have argued throughout the previous chapters, fundamental to the operational logic of biometrics. In this chapter, I examine the manner in which two technologies, electroencephalography (EEG) and functional Magnetic Resonancing Imaging (fMRI), are being developed as lie detection systems that, in the words of their makers, provide "direct," "unbiased" and "objective" methods of truth verification. EEG scans, obtained through the application of a headband with electrodes attached to the subject's scalp, measure the electric currents generated by specific brain activity. These electric signals are recorded as brain-wave graphs. Brain Fingerprinting Laboratories claim that they have developed an EEG system that reliably records whether a subject is telling the truth or is lying. No Lie MRI, Inc., deploys fMRI scans in order visually to record the specific areas of the brain that become neurologically active during the performance of particular tasks; electromagnetic radiation measures the localised brain activity through exposure to a magnetic field. Instead of brain-wave graphs, fMRI scans produce three-dimensional images of the brain, in which localised areas deemed to be "active" are often highlighted through the use of different colours.

Strictly speaking, neither of these technologies would be viewed as biometric systems, as they are not concerned with identity verification or authentication. I would argue, however, that, once situated within the complex genealogy of biometric technologies that I traced in Chapter 1, these technologies can be classified as biometric in that they use technology in order to measure (*metric*) physiological (*bios*) activities as ways of determining such moral categories as truth and lies: specific physiological markers are represented as offering *biometric signatures* of such moral categories as truth and lies. Indeed, Brain Fingerprinting's very nomenclature analogically situates this technology within traditional biometric technologies of criminal investigation, including fingerprint identification and forensic genetics. Brain Fingerprinting uses brain wave signatures as its truth verification biometric; and No Lie MRI uses images of localised neurological "hot spots" that disclose the biometric evidence as to whether or not the subject under investigation is telling the truth or lying.

In the course of this chapter, I will analyse both these technologies in the context of the claims made by their respective makers. In particular, I will focus on the sociocultural mediations that inscribe these technologies, even as these same multimodal mediations are effectively effaced in the literature that advertises the claims made by both Brain Fingerprinting and No Lie MRI. As there is already a significant body of literature dedicated to articulating the range of ethical and legal issues that these two technologies generate, I will not focus on these matters (see Uttal 2009; Moriarty 2008; Greely and Illes 2007; Tovino 2007; Garland 2004; Tancredi 2004). Rather, I will focus on the ongoing influence of biometric antecedents, such as physiognomy and phrenology, on both these technologies. Situated within these biometric genealogies, I argue that these two technologies reproduce in contemporary guise many of the concerns that were previously raised about the scientific practices of physiognomy and phrenology, including the use of biometric studies of the body in order to make diagnostic claims about a subject's moral character and their prospective behaviour and actions. As technologies fundamentally concerned with the disclosure of truth they are, once understood in Foucauldian terms, technologies that in fact *produce* the truth they claim to discover. As technologies underpinned by medico-legal protocols, both Brain Fingerprinting and No Lie MRI emerge as technologies of "demonstrative truth joined ... to scientific practice" (Foucault 2006, 236). The "demonstrative" dimensions of these technologies signify the "evidentiary" status of the truth that they claim to discover, while the authorising role of scientific practice functions to invest this truth with the gloss of unmediated objectivity.

Brain Fingerprinting as Neurognomics

Brain Fingerprinting technology is founded on the premise that a criminal's brain is permanently marked by the psychophysiological knowledge of the criminal act that the subject has perpetrated:

> The fundamental difference between the perpetrator of a crime and an innocent person is that the perpetrator, having committed the crime, has the details of the crime stored in his memory and the innocent subject does not. This is what Brain Fingerprinting testing detects scientifically, the presence or absence of specific information.
>
> (Brain Fingerprinting Laboratories)

The presence or absence of information relevant to the crime is made evident through the following process:

> In a Brain Fingerprinting test, relevant words, pictures or sounds are presented to a subject by a computer in a series with irrelevant and

> control stimuli. The brainwave responses to these stimuli are measured using a patented headband equipped with EEG sensors. The data is then analysed to determine if the relevant information is present in the subject's memory. A specific, measurable brain response known as a P300, is emitted by the brain of a subject who has the relevant information stored in the brain, but not by a subject who does not have this record in his brain.
>
> (Brain Fingerprinting Laboratories)

Brain Fingerprinting technology purports to reveal this incriminating knowledge objectively and scientifically regardless of the subject's conscious desire to hide or dissimulate their knowledge of the criminal act: "Presented with details of the crime, the guilty person cannot help but elicit an involuntary, but detectable spark of recognition in the brain. This response is automatic, so there is no way to suppress or fool the system" (Murphy 2004). The P300 has been identified as part of a brainwave response known as a MERMER (memory and encoding related multifaceted electroencephalographic response):

> A MERMER will only be emitted by the brain of the perpetrator, with details of the crime in his brain, and not by an innocent suspect who does not have this record in his brain. Find the MERMER and you have found the murderer.
>
> (Murphy 2004)

A MERMER is identified through the deployment of a "multiple-choice test for the brain":

> Three kinds of information are used to determine whether a subject has specific crime-related information in his brain:
>
> - Targets: information the subject definitely knows; this can be ensured by telling the subject before the test starts.
> - Irrelevants: information that [the] subject definitely does not know; this can be ensured by simply making up the information.
> - Probes: information relevant to the crime or situation, which the subject may or may not know.
>
> The response of the brain to information is measured using a headband with electrodes. Target information elicits a "yes" response or a MERMER. This is used as a control. Irrelevant information will not elicit a MERMER. A MERMER in response to probe stimulus indicates recognition or the presence of certain information.
>
> (Murphy 2004)

In what follows, I want to unpack the presuppositions that found and constitute Brain Fingerprinting. The central presupposition that underpins Brain Fingerprinting pivots on an understanding of technology as a mere conduit that delivers data or information in an unmediated way. This conduit understanding of technology is evidenced by the claim that "what Brain Fingerprinting testing detects scientifically [is] the presence or absence of specific information" (Brain Fingerprinting Laboratories). In other words, Brain Fingerprinting is presented as a technology that delivers information in a binary mode, as something that is either present or absent. In the conceptualisation of this neat binary, there is a failure to mark the fundamental ways in which technology cannot ever be a mere conduit for unmediated information; on the contrary, the specificity of the technology itself shapes the information. The untenability of this conduit view of Brain Fingerprinting technology is further evidenced by the following description of the process:

> The entire Brain Fingerprinting system is under computer control, including the presentation of the stimuli, recording of electrical brain activity, a mathematical data analysis algorithm that compares the responses to the three types of stimuli ["targets," "irrelevants," and "probes"] and produces a determination of "information present" or "information absent," and a statistical confidence level for this determination. At no time during the analysis do biases and interpretations by the person giving the test affect the presentation or the results of the stimulus presentation.
>
> (Brain Fingerprinting Laboratories)

The conduit understanding of technology is here reproduced in the claim that the "entire Brain Fingerprinting system is under computer control." "Computer control" is the code phrase that effectively erases any sociocultural determinants that might shape and impact on either the process or the results of the technology. Effaced from the entirety of the procedure of Brain Fingerprinting are the following sociocultural mediations: the institutional space within which the computer and subject are situated during the process; the cultural particularities and inescapable discursive mediations of the pictures, sounds or other stimuli presented by the computer; the mediating conversion of psychophysiological stimuli to mathematical data and digital algorithms, and their consequent interpretation against a statistical formula. This series of mediations can be succinctly collected under the terms "design or paradigmatic bias" and "biases of interpretation" (Greely and Illes 2007, 384).

The erasure of these fundamental mediations, mediations all inflected with their own determining "biases," discursively enables the claim that "At no time during the analysis do biases and interpretations by the person giving the test affect the presentation or the results of the stimulus presentation" (Brain Fingerprinting Laboratories). Once again, the untenability of this claim is evidenced by the simple fact that the results have already undergone an invisibilised process of fundamental mediations; furthermore, the results obtained through this series of mediations must remain unintelligible as data or information until they are actually *interpreted*, and thus mediated once again, by the person conducting the test. Represented as "human free" technology, the Brain Fingerprinting system can then be presented as bias free. As Jonathan Moreno (2006, 109) notes, "'Mind reading' technologies do not, in fact, read the mind. They are just another set of data to be interpreted contextually." Moreno (2006, 109), in addition, draws attention to the critical failure that underpins the claims of Brain Fingerprinting advocates: "Neuroscience reads brains, not minds. The mind, while completely enabled by the brain, is a totally different beast."

The fact that the detected MERMER, represented through the visual image of digitised graphs with peaks and troughs, must be interpreted in the context of brain-wave analysis situates the technology of Brain Fingerprinting within the hermeneutic tradition of physiognomics. In *Body Criticism: Imaging the Unseen in Enlightenment Art and Medicine*, Barbara Stafford (1997, 84) offers an incisive definition of physiognomics: "Physiognomics was body criticism. As corporeal connoisseurship, it diagnosed unseen spiritual qualities by scrutinizing visible traits. Since its adherents claimed privileged powers of detection, it was a somewhat sinister capability. This 'science' supposedly divined what untrained eyes could never see about a person's character." Brain Fingerprinting must be situated within this genealogy (see Chapter 1). As a contemporary digitised form of physiognomics, Brain Fingerprinting should in fact be termed *neurognomics*: its adherents claim privileged powers of detection in their reading of the contours of EEG brain wave graphs and in divining what untrained eyes cannot see. "The activity of searching inquiry," writes Stafford (1997, 84), "linked medical diagnostics to textual criticism and cerebral expertise. Symptoms ... were converted into esoteric graphic signs. These physical enigmas were indicative of hidden causes legible only to specialized interpreters" armed with the "the 'right' detectors." With Brain Fingerprinting, a technology of medical diagnostics, the EEG, is intertwined with a form of textual criticism that pivots on a test for the brain oriented around "targets," "irrelevants," and "probes." The "symptoms" that result from this brain test are converted into "esoteric graphic signs," the EEG brain wave graphs, that can only be understood by "specialised interpreters"/scientific exegetes who proceed to demonstrate their "cerebral expertise" by deploying the "right detector," a headband with electrodes, in order to disclose the brain's "hidden causes."

As with physiognomics, neurognomics is predicated on the deployment of a "consummate visual study" of corporeal signs that will "unravel th[e] 'chaos' of seeming" (Stafford 1997, 86). Neurognomics, like physiognomics, is presented in terms of a "preventative medicine against the traps of cheating appearances" (Stafford 1997, 89). Whereas in physiognomics it is the morphology of external appearances that reveals the true nature of the subject's soul, in neurognomics it is the peaks and troughs of brain waves that disclose an image of the subject's true nature. In physiognomics, the topography of a subject's true nature is mapped through a graphic process of geometric calibration and typologisation of the face: "The geometric layout of the page [in physiognomic illustrations] consisted of divisions into rectangles, circles, ovals" (Stafford 1997, 95). This geometric and mathematical computation of a subject's inner nature is elaborated, in neurognomics, through EEG graphs, where the psychophysiological detection of concealed information is topographically mapped through a series of brain-wave graphs; the gradients of these graphs reveal either information absent or information present, that is, a MERMER. The visual images of Brain Fingerprinting's identificatory brainwaves and MERMERs are strikingly prefigured in Cesare Lombroso's use of "pulse waves" and "pulse graphs" in order to identify liars and criminals (see figure 5.1). David Horn (2003, 120) observes how

> graphical instruments developed in the nineteenth century promised both a new quantification and a new kind of visibility. In this case, inscription devices transformed the study of vasomotor reactions and appeared to make newly accessible the phenomena of feeling and thinking.

In both traditional physiognomy and Brain Fingerprinting, there is operative an imperative toward "biological epitomization," (Stafford 1997, 103) in which the graphic images (the angle of a face or the contours of a brain wave that indicate a MERMER) of corporeal effects function to reveal the truth of the subject under investigation. Physiognomics, Stafford (1997, 103) argues, emerged in the context of what she terms "metric revolutions and the rise of statistics during the Enlightenment"; both these things "went hand in hand with a geometrized aesthetics." Situated in this context, the

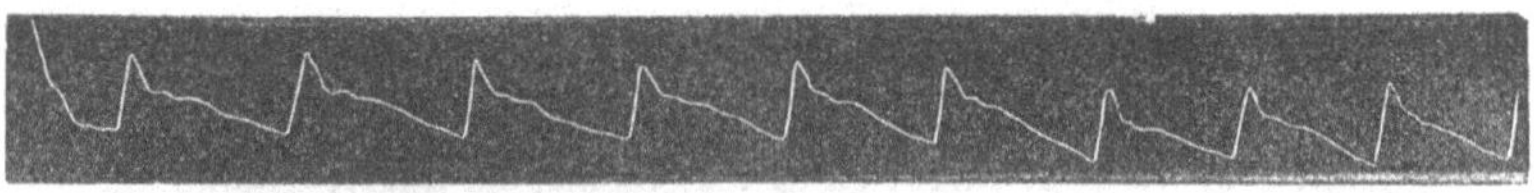

Fig. 2.

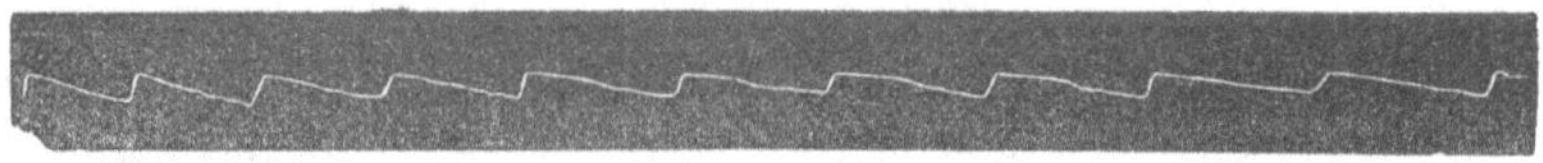

Fig. 3.

Figure 5.1 Cesare Lombroso's "pulse graphs." Cesare Lombroso, *L'Uomo Deliquente*, 1889.

genealogical affiliations between physiognomics and neurognomics might best be summarised by the fact that both are predicated on a "graphology of character" (Stafford 1997, 96) made manifest through metrical and geometrised mediations that require a "connoisseurship" in order to disclose what would otherwise remain hidden: the "truth" of the subject under investigation.

In her work, Stafford (1997, 84) identifies the central objective of physiognomics: "Cutting through density was, literally and metaphorically, a way of piercing any opaque morphology, of achieving transparent self-knowledge and the knowledge of others." Brain Fingerprinting, as a form of neurognomics, reproduces this self-same objective. The scanning technology promises to take the observer beyond the duplicities and feints of the surface of the suspect subject. Penetrating into the cerebral inner sanctum of the subject, the clinical gaze, augmented by visualising technologies, pierces the opaque morphology of skin and bone in order to disclose another type of morphology altogether, brain-waves, that will, regardless of the volition of the suspect subject, deliver up the "truth." The physiognomical presuppositions that are embedded in Brain Fingerprinting are perfectly captured in this observation on traditional physiognomics: "Physiognomical signs ... reflect an essential truth which their possessor has no power to conceal or transform" (Rivers 1994, 80).

Operative in Brain Fingerprinting's promise that its penetrative scientific gaze can disclose the truth of the subject, regardless of his or her will, is that fusion of the clinical gaze and the process of diagnostic analysis that Foucault (1975, 94) draws attention to in his work on the emergence of clinical practice:

> Analysis and the clinical gaze also have this feature in common that they compose and decompose only in order to reveal an ordering that is the natural order itself: their artifice is to operate only in the restitutive act of the original.

The mediating processes of composing and decomposing—that are at once linguistic, scopic and technological—that enable the process of diagnostic analysis are occluded by the "artifice" of claiming that what is revealed is, in fact, merely a restitution of what has always been there, silently and passively awaiting its "discovery" by the clinical gaze. Inscribed here is that "fundamental isomorphism" between the object of inquiry and the visual and linguistic forms "that circumscribe it" (Foucault 1975, 95). "The descriptive act," Foucault (1975, 95; italics original) emphasises,

> is, by right, a "seizure of being" (*un prise d'être*), and, inversely, being does not appear in symptomatic and therefore essential manifestations without offering itself to the mastery of a language that is the very speech of things ... in clinical medicine, *to be seen* and *to be spoken*

> immediately communicate in the manifest truth of the disease of which it is precisely the whole *being*. There is disease only in the element of the visible and the statable.

To a degree, the type of discursive effacement of sociocultural determinants in the actual operations of technology that I traced earlier is characteristic of the positivism that continues to inscribe western science. On another level, however, this discursive effacement can be seen to be significantly amplified by the development of digital technology that, as Jonathan Crary argues, effectively liquidates the constitutive role of the observer in the production and interpretation of digital images precisely because these images do not reproduce the mimetic representations of analogue technologies such as film and photography. The recession of mimesis from the visual plane of the observer, and its replacement by computer fabricated digital images, functions to relocate "vision to a plane severed from a human observer":

> Most of the historically important functions of the human eye are being supplanted by practices in which visual images no longer have any reference to the position of an observer in a "real," optically perceived world. If these images can be said to refer to anything, it is to millions of bits of electronic mathematical data. Increasingly, visuality will be situated on a cybernetic and electromagnetic terrain where abstract visual and linguistic elements coincide and are consumed, circulated, and exchanged globally.
>
> (Crary 1998, 1–2)

The liquidation of the mimetic effect in the visual field of digital images functions to produce the discursive illusion that imaging technologies are seemingly free of human bias. The digital image of brainwaves, for example, does not correlate to the optically perceived "real" world of the observer. However, the observer "returns," so to speak, in the after-moment of digital image production, precisely in order to make sense of and interpret the digital image. This "return" of the hermeneutic observer, as constitutive mediator/interpreter of the digital image, is precisely what is occluded by the non-mimetic status of the digital image when it is represented as structurally independent of a human observer. This "return" of the hermeneutic observer, moreover, must be seen as a structuring *a priori* that at once precedes the "reading" of the image in question even as it constitutes its conditions of cultural intelligibility. Foucault (1975, 113. Emphasis added) unpacks the operational logic of this process:

> the analytical structure is neither produced nor revealed by the picture itself; the analytical structure preceded the picture ... beneath its apparently analytical function, the picture's only role is to divide up the

> visible within an already given conceptual configuration. The task is not, therefore, one of correlation, but, purely and simply, of redistribution of what was given as a perceptible extent in a conceptual space defined in advance. It makes nothing known; at most, it makes possible *recognition*.

Articulated in this seemingly paradoxical logic is the following: that what the scientific gaze claims to "discover" or "disclose" in the purity of its objective field of perception is, structurally, predetermined to appear as such through the (effaced) analytical conditions that enable its very *re-cognition* when it is seen.

There is, in this contemporary regime of digital visuality, another paradox that must be brought into focus. As Roland Barthes (1993, 85) has argued, the analogue photograph is distinguished precisely by the fact that it represents "*what has been*": the object in the "real" world is photochemically inscribed on the surface of the film and the negative remains as the authenticator of what has been recorded. The analogue photograph is thereby invested with "an evidential force" whereby "the power of authentication exceeds the power of representation" (Barthes 1993, 89). The analogue photograph's "essence is to ratify what it represents" (Barthes 1993, 85). This is not say, of course, that the analogue photograph's representation of the evidentiary "what has been" is not itself constituted by its own complex series of mediations and fabrications (see Mitchell 1998). The digital image, on the other hand, as computer-fabricated visual artefact, cannot reproduce the mimetic-evidentiary effect of analogue film and photographs: the representation of "what has been" is wholly constituted by a series of mathematical algorithms that may bear no effective relation to the object represented, as the image may be the result of a series of fabrications "from found files, disk litter, the detritus of cyberspace. Digital imagers give meaning and value to computational readymades by appropriation, transformation, reprocessing, and recombination; we have entered the age of electrobricollage" (Mitchell 1998, 6).

Despite this disjunctive rupture between the two media, the digital image is still mobilised as a visual artefact invested with an incontrovertible evidentiary status: Brain Fingerprinting Laboratories have presented in the court of law images of brain wave graphs in the Terry Harrington case. In her discussion of Brain Fingerprinting in the context of the Terry Harrington case, in which Harrington was charged with murdering a security guard, Marina Murphy (2004) describes the digital images that were presented in court as having operated as "Harrington's get out of jail card," supplying the necessary evidence to prove his innocence. "The judge ruled," Murphy (2004) writes, "that the B[rain] F[ingerprinting] test results 'meet the legal standards for admissibility in court for scientific evidence and the murder conviction was reversed in 2003." However, as Jane Moriarty (2008, 35) notes in her analysis of the use of Brain Fingerprinting in U.S. courts, "'brain fingerprinting' has been mentioned in a few U.S. cases, but has not been relied on by any court to date."

In his tracking of the way in which digital and electromagnetic visual technologies are "becoming the dominant modes of visualization according to which primary social processes and institutions function," Crary (1998, 2) draws attention to the manner in which they are "intertwined with the needs of global information industries and with the expanding requirements of medical, military, and police hierarchies." Brain Fingerprinting Laboratories delineate, in their prospectus, the criminal-justice uses of Brain Fingerprinting, adding that, in the context of the war on terror, the technology has important national security applications.

In the conceptual schema of Brain Fingerprinting, the brain is understood to be a mere recording device and repository of past and intended acts and thoughts. Despite the fact that digital technologies are used to access and process the brain's memories and thoughts, Brain Fingerprinting actually relies on an analogue model of the brain that, like photochemical film, proceeds to record, etch and store its information. When confronted, analogically, by a relevant picture or word equivalent of a thought or memory, the brain automatically and involuntarily produces the fabled MERMER. This MERMER is detected by an EEG and is then transmuted into a visual artefact of brainwaves on the computer screen. There are a number of fundamental problems in this representation of the brain's operation. In the first instance, this analogue, and decidedly static, model of the brain is untenable when situated in contemporary neuroscience studies of the brain that proceed to represent the brain as dynamic, plastic and complexly interconnected across different neurological levels. As Alloway and Pritchard (2007, 379) note:

> No single neural structure can mediate complex processes such as motivation, memory, or emotion. The autonomic, cognitive, and behavioural responses associated with these processes are mediated by a distributed system of interacting neural structures that include the amygdaloid complex, the hypothalamus, the basal forebrain, the nucleus accumbens and the ventral striatum, the prefrontal and angulate cortices, and a collection of somatic and autonomic motor nuclei in the brainstem.

Situated in this neuroscience understanding of the brain as complexly interconnected in its operations, the evidentiary singularity of the MERMER, as detected by the EEG and represented as supplying *the* identificatory evidence of a past or intended criminal act, remains dubious. Indeed, in his "Brain Dynamics: Modelling the Whole Brain in Action," Jim Wright (2000, 155, 141–42), in presenting a view of brain dynamics "as processes embedded within each other at multiple levels," proceeds to question the representation of the EEG as a "cognitive mirror" by arguing that EEGs can only be viewed as offering "indistinct" representations of the operations of the brain: "The EEG has often been dismissed as an epiphenomenon having little if any strong relationship to more fundamental events in the brain, whatever these are." Furthermore, the mimetic concept metaphor of the brain as "cognitive mirror" reproduces a reductive understanding of the brain in which its operations are premised, once again, on the simplistic model of an analogue-recording device. In the context of the complex interconnectedness and non-linearity of brain functions, William Uttal (2009, 35) argues that EEG scans

> must be considered to be very 'blunt' tools. These signals are the accumulation of the responses of literally billions of neurons. Thus, they throw away all of the critical and detailed information about the activity of the vast network of neurons that most likely account for mental processes.

Furthermore, neuroscientists have drawn attention to the fact that MERMERs can actually be voluntarily produced by a subject undergoing an EEG scan:

> There's also evidence that people can produce P300 waves by deliberately thinking about stimuli that have never taken place except in their imagination, such as a group of students in one study who produced a bump after they were told to think of their instructor slapping them. Terrorist's brains might show that they're well aware of terror targets and methods, but so might the brains of journalists and intelligence experts—and teenage boys who find jihadist web sites titillating.
>
> (Moreno 2006, 100)

The problematic nature of touting Brain Fingerprinting as a foolproof system for detecting evidence of criminal acts becomes evident when situated in the context of a case that unfolded at the University of Nottingham in the United Kingdom, where a "graduate student at the School of Politics and International Relations and an administrative member of the staff at the Department of Engineering were arrested by armed police under the Terrorism Act of 2000":

> Their alleged "crime" was that the graduate student had downloaded an al-Qaeda training manual from a US government website for research purposes, as he's writing his MA dissertation on Islamic extremism and international terrorist networks. He had sent this to his friend in the Department of Engineering for printing. The printed material had been spotted by other staff and reported to the University authorities who passed on the information to the police. The two were arrested by armed police on May 14 [2008] and held for six days without charge, before being released without charge on May 20. During the six days

> they were imprisoned, the men had their homes raided and their families harassed by the police. It is worth noting that in talking to one of my colleagues, a police officer remarked that the incident would never have occurred if the persons involved had been "blonde, Swedish PhD students" (the two men were of British-Pakistani and Algerian backgrounds respectively).
>
> (Nilsen 2008)

This case is of particular interest in questioning the evidentiary value of Brain Fingerprinting for criminal prosecution. In the first instance, it brings into clear focus the manner in which visual regimes of racial profiling place subjects with particular phenotypical features under suspicion before the fact of their actually having committed a crime. The biological racism that underpins regimes of racial profiling is still actively reproduced in the forensic field of criminal justice (see Holbert and Rose 2004). For example, in *Forensic Aspects of Speech Patterns: Voice Prints, Speaker Profiling, Lie and Intoxication Detection*, Tanner and Tanner (2004, 47) proceed to offer this definition of race in the context of their discussion of racial profiling: "A person's race is a biological determinant where skin color, body type, facial features and other physical attributes are passed genetically from generation to another." In other words, race here is seen to be biologically immanent and, as such, it functions as a type of *a priori* that precedes, in its biological and genetic positivity, the analytic procedures and discursive schemata of the forensic scientist. Stuart Hall (1993, 297–98; italics original) offers a trenchant critique of the ongoing biological use of race:

> contrary to a widespread belief—race is not a biological or genetic category with any scientific validity. There are different genetic strains and "pools," but they are as widely dispersed *within* what are called "races" as they are *between* one "race" and another. Genetic difference—the last refuge of racist ideologies—cannot be used to distinguish one people from another. Race is a *discursive* not a biological category. That is to say, it is the organising category of those ways of speaking, systems of representation, and social practices (discourses) which utilize a loose, often unspecified set of differences in physical characteristics—skin colour, hair texture, physical and bodily features etc.—as *symbolic markers* in order to differentiate one socially from another.

Situated within the operational logic of Brain Fingerprinting, and its dubious methodology of "multiple-choice test for the brain" in which targets, irrelevants and probes are used, if these two University of Nottingham "terror suspects" had undergone an EEG brain scan in which al-Qaeda print or visual texts had been used as "event-related potentials" to disclose knowledge of al-Qaeda terrorist practices, an "evidentiary" MERMER is precisely what would have been produced, incriminating the two subjects despite the fact that they were merely conducting scholarly work on the operations of a terrorist organisation.

I draw attention to this case in the context of the national security applications that Brain Fingerprinting Laboratories advertise on their website:

> In a terrorist act, evidence such as fingerprints or DNA may not be available, but the brain of the perpetrator is always there—planning, executing, and recording the crime. There are memories of the crime stored in the brain of the perpetrator and in the brains of those who helped plan the crime. Brain Fingerprinting Laboratories technology can detect these records stored in the brain and help identify trained terrorists before they strike, including those that are in long-term "sleeper" cells.

The aforementioned case begs the question as to how Brain Fingerprinting technology can effectively differentiate between "memories of the crime" that result either because one has been an actor in or observer of the crime and those memories that are the result of having read a text or viewed a video of the crime. Brain Fingerprinting technology is predicated on an entirely positivist understanding of memory and the operations of the brain. In its conceptual schema, the brain merely records events and these then become available to an EEG scan as evidentiary traces of what has taken place: "The concept behind the main product of Brain Fingerprinting Laboratories, Inc., is to identify memory *traces* through brain wave responses" (Moreno 2006,104. Emphasis added). Building on my Derridean critique of biometric technologies' dependence on the metaphysics of presence in order to combat imposture and fraud, I want to proceed to unpack the problematics that arise from conceptualising, as Brain Fingerprinting Laboratories proceed to do, the figure of the *trace* in positivistic terms.

As I explained in some detail in Chapter 4, Derrida has demonstrated how the identity of a subject or sign is never self-identical nor fully present to itself. To recap briefly, the identity of a subject or sign is constituted, in Derridean terms, through the operations of *différance*: "every concept is inscribed in a chain or in a system within which it refers to the other, to other concepts, by means of the systematic play of difference" (Derrida 1986, 11). This differential relation is also constituted by a complex system of deferrals. The identity of a sign, then, is merely a trace whose identity is constituted "by means of this very relation to what it is not" (Derrida 1986, 13). It is thus, Derrida (1986, 11) argues, that: "in language there are only differences *without positive terms*." There are no positive terms because the sign is only ever constituted as a trace that is divided in and of itself by all the terms from which it differs and which it simultaneously defers.

I want to transpose this Derridean concept of the trace to a critique of the "traces" of memory that Brain Fingerprinting systems purport to detect and capture with their EEG brain wave scans. Situated within Derrida's critique of empirico-positivist understandings of identity, the traces of memory, "these records stored in the brain," must be viewed as virtual fabrications that elide the complex differential relations that inscribe and constitute them. These traces of memory cannot be captured and reproduced in any self-identical manner. The graphic traces that represent them, the visual brainwave graphs, are static simplifications of complex and plastic neurological operations—that include biochemical reactions, synaptic firings and connections, and the interacting of a cluster of neural structures—that cannot be reduced to a single neurological artefact, that is, an EEG graph. The EEG brainwave graph, as an evidentiary trace of memory, is already divided in and of itself by the complex ensemble of neurological differentials and deferrals that constitute its very conditions of possibility. In other words, the virtual representation of the trace of memory, the brainwave graph, always remains nothing more than a virtual representation that can never be self-identical to what it purports to represent. Precisely because it is a visual sign, it is already enmeshed within differential networks (neural, discursive and semiotic) that always already preclude the possibility for it to be a purely positive/positivistic term. Indeed, if the logic of the trace is predicated on its constitutive relation to what it is not and what is not "there," then the trace of memory that Brain Fingerprinting purports to capture and record must also be viewed as "a vestigial 'memory' of non-presence" (Critchley 1992, 75).

What is problematised, then, through this deconstruction of Brain Fingerprinting's reliance on the memory trace as positivistic artefact, is the technology's system of conceptuality, a system predicated on the binary of "'information present' or 'information absent'" (Brain Fingerprinting Laboratories). It is the system's polar binary of presence/absence that becomes untenable when critiqued via Derrida's mobilisation of *différance* and the trace. In Derrida's deconstructive schema, nothing can be simply and reductively located and fixed within this disjunctive binary. This polarised, positivistic binary (present/absent) can only be produced and maintained through the elision of the complex network of differences and deferrals, as outlined earlier, that enable its very conditions of intelligibility.

I want to elaborate on the spectral ramifications of virtual memories and vestigial memories of non-presence by drawing attention to the biopolitical presuppositions that inform Brain Fingerprinting's understanding of the brain. Operative in Brain Fingerprinting Laboratories' claim that memories of crime can simply be detected as either absent or present is a normative conceptualisation of the brain that fails to account for brains afflicted with various forms of mental illness that produce real and embodied neurological effects that problematise the simple absent/present binary. Nowhere in Brain Fingerprinting's literature is the question of mental illness broached or discussed. How would Brain Fingerprinting differentiate, for example, between psychotic states, where the brain produces its own psychological "realities," and all of their attendant physiological effects, "realities" which, for the subject in question, are neither fabrications nor lies but vivid actualities? How, furthermore, would Brain Fingerprinting technologies deal with subjects who, due to post-traumatic stress, have psychophysiologically "wiped" intolerable memories as a way of surviving their trauma? In his critique of Brain Fingerprinting systems, Moreno (2006, 106) writes:

> As a forensic device, critics of brain fingerprinting note that the test really measures whether the subject is familiar or unfamiliar with a crime scene, not deception. In the only published study on the subject, the ERP [Event-Related Potentials] detector didn't do much better than guesswork. Also, if the subject has forgotten an event or if he or she is mentally ill, the response could be altered.

Furthermore, as Margaret Talbot (2007, 58) argues in "Duped? Can Brain Scans Uncover Lies?", "researchers have since [the development of Brain Fingerprinting] noted a big drawback: it's impossible to distinguish between brain signals produced by actual memories and those produced by imagined memories—as in a made-up alibi."

Despite the inordinate claims made by the makers of Brain Fingerprinting concerning the technology's ability to detect truth from lies, there have been virtually no independent scientific studies published that support their claims. "One stumbling block to validating the P300," writes Moreno (2006, 106), "is the fact that the techniques used to analyse brain fingerprinting results are proprietary, so they can't be analysed by independent scientists." This is a view articulated in many of the reviews of lie-detection technologies. Wolpe et al. (2005, 45), for example, offer this sceptical conclusion: "Most of the data is not published in peer-reviewed literature. Thus, the true accuracy, validity, and relevance of this method to any real-world applications must be deemed unknown by any modern scientific standard." In his critique of Brain Fingerprinting technologies, William Uttal (2001, 91) describes the enterprise in terms of the "tendency to concretise and reify the ephemeral"—in other words, all that Brain Fingerprinting can effectively claim to achieve is the concretising of transient brain waves into fixed graphs.

No Lie MRI as Digital Phrenology: The Neurological Locus of Lies and the Digital Topography of Neural Locationism

Another technology that parallels the claims made by Brain Fingerprinting is No Lie MRI. No Lie MRI uses functional Magnetic Resonance Imaging in order to reveal whether or not a subject is lying or telling the truth. This imaging technology captures areas of the brain that appear to be activated during particular cognitive operations by measuring changes in blood flow. Areas of the brain that are seen as active are indicated by volumes of higher blood flow. fMRI imaging technology represents this area of neurological activity through voxels or three-dimensional pixels. Although using an entirely different technology, the claims made by the manufacturers of No Lie MRI are similar to those of Brain Fingerprinting: "No Lie MRI, Inc. provides unbiased methods for the detection of deception and other information stored in the brain. The technology used by No Lie MRI represents the first and only direct measure of truth verification and lie detection in human history!" (No Lie MRI). Once again, the conduit metaphor informs No Lie MRI's understanding of the relation between technology and information. The deployment of the terms "unbiased" and "direct measure" underscores the notion that neither the technology nor the observer/interpreter mediates the information that is being represented; rather, it is delivered in a "direct" and "unmediated" manner. José van Dijck (2005, 5) succinctly identifies the paradox that is operative in this claim: "Mediated bodies are intricately interlinked with the ideal of transparency." The fact that bodies are, in the context of these technologies, mediated by the very technologies designed to deliver the "objective" information is effaced precisely by the conduit view of technology. The conduit view of technology invisibilises the constitutive effects of the technology, on both the body and the information it is made to deliver, through the ruse of *transparency*. Through the deployment of the ruse of transparency, both the body and its information are delivered up to the scientific gaze in a seemingly unmediated manner. This governing metaphor of transparency operates on two levels within the discursive operations of western science: it invisibilises both the structuring effects of the technology and the mediating role of the observer, effectively effacing the ways in which "seeing is intervening" (Ian Hacking cited in van Dijck 2005, 8).

Inscribed in No Lie MRI's view of information, the body and technology is what Eugene Thacker (2003, 85) terms the "classical theory of information": "The medium of information (to be distinguished from the message and from information) is transparent with respect to information, so that information is taken to be abstracted and self-identical across different media, or across different technological platforms." No Lie MRI is critically dependent on "algorithms to automatically analyze functional Magnetic Resonance Imaging." Yet the mediative effects of the visual technology, the algorithms and the scientific interpreter that at once encode and transcribe the materiality of the body into digital information, must be abstracted in order to evidence the claim that the technology is "observer independent (objective)" (No Lie MRI). "The point is not," Katherine Hayles (1999, 13) explains, "that abstracting information from a material base is an imaginary act but also, and more fundamentally, that conceiving of information as a thing separate from the medium instantiating it is a prior act that constructs a holistic phenomenon as an information/matter duality."

This information/matter duality is what enables the fiction that No Lie MRI "objectively measures intent, prior knowledge, and deception" of the subject under investigation.

No Lie MRI claims to have isolated specific locations in the brain that are activated whenever a lie is told. In the coloured images that illustrate the No Lie MRI website, three images of the brain (right and left sides and an anterior view) are shown to identify the areas active when a lie is told. These areas, the medial frontal gyrus, inferior parietal lobule and inferior frontal gyrus, are nominated as the identificatory "hot spots" of lies and deception. The use of technology in order reveal the embodied moral qualities of a subject has, as I discussed in Chapter 1, a long and varied history that includes the visual gallery of physiognomics, phrenology, fingerprint biometrics, the use of craniometers and topographies of the skull in order to reveal the innate dispositions of "born criminals," and the use of X-rays for socio-moral purposes. "It was a popular tenet in the early twentieth century," van Dijck (2005, 89) notes, "that X-rays could show more than bones in a living body; the cathode rays could purportedly expose a person's naked character and uncover his or her deepest passions." In her detailed mapping of the uses and abuses of X-ray, Lisa Cartwright (1995, 4) has brought into sharp focus the "institutional practice of using photography to classify, to diagnose, and to maintain control over subjects deemed criminal, mentally ill, or otherwise aberrant."

On a number of levels, No Lie MRI's claims effectively situate this new technology within the genealogy of phrenology: first, in its use of neurological locationism; second, in its deployment of a criminal-moral diagnostics of corporeal matter; and, third, in its use of visual images in order to evidence its claims. In his *The New Phrenology: The Limits of Localizing Cognitive Processes in the Brain* (2001), William Uttal stages a detailed critique of the reinvention of phrenology in contemporary digital technologies. As outlined in the work of Gall and Spurzheim (see Chapter 1), phrenology purported to be able to locate specific psychological and moral functions in particular locations of the brain. As Uttal (2009, 32; 2001, 103) argues, although phrenology's claims have long been ridiculed as nothing other than the product of a pseudoscience, the concept of brain locationism

> persists to this day in modern forms. Specifically it is the conceptual foundation of much of the current work done on brain imaging and cognitive neuroscience in general: approaches that also assume that different sites of activation are associated with different mental states.

The assignment of specific brain locations to the production of truth and lies at once reproduces traditional phrenology's locationist claims, even as it transmutes this phrenological tradition into an updated science reliant on the latest imaging technology (fMRI).

No Lie MRI "evidences" its neurological "hot spots" for lying and deceit by colour-coding its pixelated images of the brain: red for lies and blue for truth. Presented as empirical traces of targeted neurological activity, the digital colouration of "hot spots" is, in fact, equivalent to the manipulative effects of colouring-in images of the brain. William Mitchell (1998, 6) illustrates the correlation between painting and digital imaging by discussing the way in which a digital image can be produced by employing "the cursor of an interactive computer-graphics system to select pixels and assign arbitrarily chosen values to them: this makes digital imaging seem like electronic painting, and indeed some computer programs for this purpose are commonly known as 'paint' systems." Uttal (2001, 168) highlights the artefactual, constructed and arbitrary status of these digitally "painted" images:

> The problem is that the assignment of colours to various levels of activity is completely arbitrary. It is up to the experimenter to decide whether any of the various levels should be colored and what colors should be used and thus what significance should be attached to the different scores produced by subtraction of the control and experimental images. A conservative assignment could hide localized activity and a reckless one suggest unique localizations that are entirely artefactual.

Even as the choice of colours used to highlight localised areas of neurological activity is arbitrary, the semiotic charge of specific colours is not. Colours are social signs that are inscribed with cultural meaning. The semiotics of colour encodes moral, aesthetic, religious, political and ideological values. In their analysis of "colour as a semiotic mode," Gunther Kress and Theo Van Leeuwen (2002, 358) examine the way in which colour is semiotically invested with multimodal meanings and values, including the way in which colours are used in various media in order to "express *character*." No Lie MRI's use of colour in their digital images attests to the use of colour in order to indicate moral and character values. Their use of blue for truth is underpinned by the semiotic significations of this colour, which include moral authority, legitimacy and ethical behaviour (as exemplified by the way in which blue is the colour of the United Nation's "corporate identity" and "the color of the helmets of the UN-troops" [Seele 2007, 9]). Conversely, red carries the semiotic charge of someone who is dangerous, intemperate and unreliable (Kress and Van Leeuwen 2002, 343). Despite the fact that No Lie MRI would argue that it is merely deploying these two colours in order to illustrate distinct areas of localised brain activation, the particular use of colours reproduces a series of semiotic meanings that, in turn, metaphorically colour the viewer's reading of the image at hand. At once arbitrary and semiotically overdetermined, No Lie MRI's colour painting of their digital images of the brain establishes subject and reader positions in the consumption of their images. An optics is never simply about seeing; "an optics," Haraway (1991, 192) underscores, "is a politics of positioning." This colour painting of fMRI images also effectively and arbitrarily occludes, as Uttal demonstrates, the many other areas of brain activity that might also have been active but that have been edited out of the picture due to the circumscribed acquisition and algorithmic protocols (Uttal 2001, 171; see also Greely and Illes 2007, 4).

In the context of No Lie MRI's claims to have isolated the brain "hot spots" indicative of lies and deception, much of the scientific literature problematises and debunks neuroreductionist claims of brain locationism by arguing for understandings of the brain that take into account the complex interconnectivity and multi-levelled embeddedness of its operations. In this literature, the brain emerges as a "complex, interactive, and nonlinear mechanism," a "system of nodes and loci that are *collectively* responsible for behaviour" (Uttal 2001, 19, 18). In their overview of human brain imaging technologies, Gordon et al. (2000, 243) argue that:

> A crucial factor in the appropriate interpretation of brain imaging technology data is to be cognizant of the limitations of reductionism, particularly in the face of the complexities of the brain as an adaptive dynamical system. As we have re-iterated, there is *no* one-to-one relationship between anatomical site-brain function-behaviour. Much brain imaging data is presented in localizationist and reductionist terms. Beware of brain scientists bearing localizationist ("hot spots") computerized phrenology color images of the brain ... Interpretations of all brain images should also be coupled with an understanding of anatomical connectivity, specific mechanisms of brain electro-chemistry and overall organization of brain function—which goes way beyond the platitudes that multiple regions are connected and interact.

No Lie MRI's images of "hot spots," that isolate lies and deception, reproduce in contemporary digital form phrenology's brain maps with their topography of situated cognitive and moral attributes. In the words of one of the leading exponents of nineteenth-century phrenology, George Combe (and Spurzheim 2007 [1834], 19), "particular mental powers are indicated by particular configurations of the head." In No Lie MRI, phrenology's topography of cranial bumps has been replaced by three-dimensional images of the brain in which the designated "hot spots" have been colour-coded for easy identification. Phrenology's taxonomy of the "organs" of the brain produced a series of orders and genera that mapped such categories as "feelings," "sentiments," "intellectual, perceptive and reflective faculties" onto particular areas of the brain. According to Combe (and Spurzheim 2007 [1834], 131), lying was located in the "organ of secretiveness," which was "situated in the middle of the lateral region of the head, immediately above that of destructiveness." Unsurprisingly, Combe (and Spurzheim 2007 [1834], 56) claimed to have found the "organ of secretiveness" to be "large in a great number of habitual thieves." Moreover, in one of the numerous racialising moments that mark Combe's (2008 [1834], 45) text, the criminal propensity of this organ was also "large in ... American Indians ... and Hindoos."

Situated in the genealogy of phrenology, the red "hot spots" reproduced in No Lie MRI images of the brain function as digital proxies of criminality or, even more problematically, as proxies of an intended or incipient criminality, as No Lie MRI claims that it can also "objectively measure intent." Similar, then, to the claims made by Project Hostile Intent, as discussed in Chapter 3, No Lie MRI emerges as another form of digital divination, in which certain neural "hot spots" are interpreted as magically disclosing prospective criminal activities of a subject in advance of the fact of having actually committed any offence. In their review of emerging neurotechnologies for lie-detection, Wolpe et al. (2005a, W5), rather than use the word "divination," deploy the term "mind reading" to debunk the extravagant claims of these emerging technologies:

> One important confusion, which is reflected in public discussions of brain imaging and is implied in some of the commentaries, is between "mind reading" and a specific brain activity-behavior correlation, as exemplified in the lie-detection procedures we describe … Although mind-reading is a loose concept, in relation to neurotechnologies it suggests an ability to identify someone's thoughts by analyzing an fMRI or EEG recording without correlating it to a formal query procedure … At present, we have no ability to mind read in that sense and have no prospect of doing so in the foreseeable future.

I use the word "divination" as it accurately captures the manner in which nineteenth-century phrenology operated, in Van Wyhe's (2004, 15, 16) words, as a form of "diagnostic cum character divination": "The deep order of nature had to be visible, and individual observations revealed that order." Van Wyhe (2004, 58, 59), in his history of nineteenth-century phrenology, describes how phrenology was positioned as "knowing about others or revealing their secrets": "Phrenology provided the authority of a scientific seer." In the contemporary context, the No Lie MRI scientist is positioned as a divining seer who is capable of exposing, through an fMRI image, an event that has not actually taken place but has merely been "intentioned." Extraordinarily, No Lie MRI purports to present an unproblematic commensuration between an image of localised biochemical neural activity and an act that might take place in the future. The complex phenomenality of a future event is simply reduced to a localised glimmer of colour-coded cerebral photo-luminescence. The colour-coded digital "blob" of the inherent criminal intent is presented as graphic evidence of what has not actually taken place but might perhaps take place. "What credence," asks Uttal (2001, 146), "can be given to the quest to localize such phantoms?"

No Lie MRI's images of (criminal) "intent" purportedly evidence the machining of such "phantoms," specifically, the spectres of criminal acts to come: their futurity is apparently fixed and arrested by the magical combination of algorithms, visual scanners and the divining role of the scientist-hermeneut. The machining of the "lie" is accomplished through a series of mediations: algorithmic, digital, visual and hermeneutic; yet it is precisely these interconnected levels of mediation that must be invisibilised and effaced in No Lie MRI's literature in order to deliver the claims that the technology "provides unbiased methods for the detection of deception" and that it "represents the first and only direct measure of truth verification and lie detection in human history!" (No Lie MRI). This is a claim that, in the stakes of competing hyperbole, is perhaps only outdone by the rival MRI lie detection company, Cephos Corporation: "All 'readings' [of the fMRI images] are performed by computers and thus no human interpretations are required" (Cephos Corp). The effective liquidation of the constitutive role of the interpreter here signals the "presumption of a transcendental epistemic locus" (Babich 1994, 88), a presumption that positions the interpreter as wholly extrinsic to the scientific operations in question.

No Lie MCPS claim that it provides a "direct measure" of truth and lie detection re-inscribes this science back into the positivist domain of traditional phrenology: "Nature and positive facts," declared Combe (and Spurzheim 2007 [1834], 119), "are the only authority which Phrenologists acknowledge." The "positive facts" of neurological "hot spots," represented in the No Lie MRI literature as objective indicators of lying, effectively transmute the technological, algorithmic, and hermeneutical processes necessary for the production of such cultural artefacts as fMRI images into organic, unmediated and self-evident "nature." "Nature," as unmediated *bios*, reveals, through fMRI images, the morphology of past and future criminal acts and thoughts. As with phrenology, the localised status of these neurological criminal signs establishes a contemporary digital version of phrenological biotypologies of criminals. Operative in contemporary digital phrenology is what Brent Garland (2004, 34) aptly terms "neuroscientific essentialism": "the idea that one's essence is in one's mind/brain."

No Lie MRI emerges, in this context, as exhaustive in its capacity to detect lie and deceit as it can locate the neurological topology of both past and future criminal acts. No Lie MRI can, unequivocally, be seen to reproduce the prophylactic claims of nineteenth-century phrenology—it has simply replaced phrenology's cranial or cephalic index of criminality with a digitised neurological one. "We live in a more subtle century, but the basic arguments never seem to change," writes Gould (1996, 173): "The crudities of the cranial index have given way to the complexity of intelligence testing. The signs of the innate criminal are no longer sought in stigmata of gross anatomy, but in the twentieth-century criteria: genes and fine structure of the brain." The anatomy of the "born" or incipient criminal is evidenced by the techno-visualisation of an act that has not taken place but is predetermined, congenitally, to unfold in the future. As discussed in my analysis of Project Hostile Intent, the animating logic of these systems is predicated on the future anterior: in the future, as evidenced by the digital image of a subject's brain, a criminal act will always already have taken place. The arresting image of biochemical neurological activity, at once localised and colour-coded, discloses the "truth" despite the subject. Instantiating a radical split between the subject and her or his body, as discussed in Chapter 3, the body of the subject in question is scripted as involuntarily caught in a type of confessional mode that betrays the inner most secrets and thoughts of the subject in question. The scanned and imaged body will profess lies, guilt or criminal intent physiologically through the localisation of designated "hot spots" so that the target subjects bespeak their criminality despite themselves. The nineteenth-century genealogical roots of this seemingly cutting-edge digital technology can be further amplified when situated in Foucault's (2006, 304) analysis of the embedding of psychiatric power within scientific technologies of demonstrative truth: "the neurologist says: Obey my orders, but keep quiet, and your body will answer for you by giving responses that, because I am a doctor, I alone will be able to decipher and analyze in terms of truth."

What is operative here, as outlined earlier in my discussion of Project Hostile Intent as a biopolitical technology, is a regime of truth through extortion, where a so-called non-invasive technology, fMRI, invades the brain of the subject to extort what has not been voluntarily offered. No Lie MRI's claim to be "objective," because it is enabled by algorithms that "automatically analyze functional Magnetic Resonance Imaging," echoes the claims made by Lombroso for the nineteenth-century index-craniograph. Lombroso, as Horn (2003, 84) explains, describes this instrument as "'automatically giving' an index as a relation of two cranial diameters." As with No Lie MRI, the mediative effects of the technology, the algorithms and the scientific interpreter that at once encode and transcribe the materiality of the body into information, must be abstracted in order to evidence the claim that the technology is "observer independent (objective)." Remarking on the index-craniograph, Horn (2003, 84) observes: "Such devices, in fact, appeared to read or record indices directly from the body and in an unmediated way; in a sense, it was the criminal body itself that produced the index, telling the story of its own dangerousness. Ironically, the scientist, who was elsewhere so insistent on his special ability to read the body, was in the act of collecting an index made to withdraw from view."

Uttal's trenchant critique of brain imaging lie detection technologies has been based on an exhaustive meta-review of the scientific literature. In his conclusions, Uttal (2009, 78) argues that "it becomes clear that no single region of the brain was found to consistently indicate deception in this group of studies ... Even worse was that none of the 16 studies identified the same localized region or pattern of regions as being associated with lying!" (see also Greely and Illes 2007, 16). The lack of consistency across all these studies (in terms of determining the neurological location of lying) functions to undermine No Lie MRI's claim that its technology is a "product that objectively measures intent, prior knowledge, and deception" (No Lie MRI). Uttal (2009, 100) emphasises the untenability of this claim when he outlines his "inescapable conclusions" at the end of his meta-review of the literature:

1. There is no evidence of a single localized area associated with deception. Peaks of activation are reported throughout the brain in what seems to be a virtually random manner by this group of experiments.
2. There is, furthermore, no evidence of any reliable broad pattern of activations associated with deception. Statistically significant responses occur scattered about nearly all portions of the brain in a manner that varies from report to report.

The question of "where" that drives the contemporary phrenological project of neurological locationism and its quest to isolate and demarcate the topography of lying is shown to be, as Uttal (2009, 104) sardonically concludes, "almost everywhere!" Complicating things still further is the issue that subjects undergoing the fMRI lie detection tests can deploy, unbeknownst to the scientist conducting the experiment, any number of countermeasures that can distort and invalidate findings:

> Simple movements of the tongue or jaw will make fMRI scans unreadable. Movements of other muscles will introduce new areas of brain activation, muddying the underlying picture. Even less visibly, simply thinking about other things during a task may activate other brain regions in ways that interfere with the lie-detection paradigm.
>
> (Greely and Illes 2007, 403)

On the Rhetoricity of Lies and Truth and the Art of Science

Aside from the widely differing empirical evidence that renders the correlation of delimited neurological locations with lies untenable, and the possible use of countermeasures in order to distort results, the fundamental question that encompasses all lie-detection technologies is: how does one define a lie? The magnitude of the problems posed by this question is lucidly outlined by Talbot (2007, 59):

> The word "lie" is so broad that it's hard to imagine that any test, even one that probes the brain, could detect all forms of deceit: small, polite lies; big, brazen, self-aggrandizing lies; lies to protect or enchant our children; lies that we don't really acknowledge to ourselves as lies; complicated alibis that we spend days rehearsing. Certainly, it's hard to imagine that all these lies will bear the identical neural signature. In their degrees of sophistication and detail, their moral weight, their emotional valence, lies are as varied as the people who tell them.

Greely and Illes (2007, 405) also pose this confounding spectre: that "a well-memorized lie may not activate those additional regions [localised to lies] and may look like a truth." Uttal brings into critical focus the problematics of defining, in any watertight and strictly categorical manner, the concept of the "lie." In his discussion, he tracks the difficulty in defining what he terms "a psychological construct such as the word 'lie'" (Uttal 2009, 79). Rather than define lies in terms of psychological constructs, I would approach the problem from a Foucauldian perspective and argue that both truth and lies are *discursive* constructs. As such, there can never be a trans-historical and universal definition of a lie; rather, one can begin to analyse the contingent status of both truth and lies as determined precisely by regimes of knowledge and power. In Foucauldian terms, discourses function to determine, define and regulate their objects of inquiry; discourses are invested with relations of knowledge/power because they are invariably grounded and authorised by institutional sites and bodies (for example, science or law). A discourse thus produces and reproduces the values, beliefs, and the preferred or dominant meanings of an institution or authority. The institution and its figures of authority (for example, scientists or judges) maintain and police a discourse's regulative function. The relations of power that inscribe discourses also effectively determine the positions that its subjects may occupy, that is, it organises the speaker and listener, for example, in terms of regulated and reproducible relations of power.

In his genealogical examinations of such categories as sexuality, madness and criminality, Foucault tracks the discursive mutability of these seemingly self-evident and ahistorical categories. In his *Order of Things* (1973), for example, he discloses the discursive conditions of intelligibility that rendered certain cultural practices as scientific in their time, while they are merely viewed as pseudo-scientific or as completely unscientific in the contemporary context. The discursive status of scientific truth, its historical contingency and mutability, is precisely what emerges in the study of the literature of neurological science. In his meta-reviews of the relevant scientific literature on the brain, Uttal (2001, 126) has drawn attention to the manner in which claims based on neurological locationism are marked by their historical contingency and transience: "It is difficult indeed to localize a process or a function in a particular part of the brain when that process or function is so ephemeral that it does not last even a single generation."

Many of the scientific studies on neuroimaging lie detection technologies articulate an informed scepticism about the technologies' inordinate claims by focusing on the complexities of defining the criteria that underpin the category of the lie as used in methods of lie detection. Donald Kennedy (2005, 19) argues that these technologies amount to little more than examples of "post-modern phrenology." Wolpe et al. (2005, 45) bring into focus the way in which designated areas of neurological activation detected during lie detection tests are not necessarily exposing empirical evidence of lies as such:

> None of the new imaging technologies actually detect "lies." Techniques such as fMRI, P300 electrophysiology, or "brain fingerprinting" detect

> physiological changes, such as blood flow or increased electrical activity in the regions of the brain that might be activated by the act of deception per se, or by the visual or psychological salience of a particular test item to the individual being tested. Separation of a deception-related signal from the host of potentially confounding signals is a complicated matter, and depends on the careful construction of the deception task rather than the measurement technology.

Wolpe et al. here underscore the complex task of separating out all the connected and distributed neurological activities that effectively problematise reductionist localisation experiments and that fail to account for experiments that nominate the same designated neurological "hot spots" for a cluster of other activities. As Langleben et al. (2005, 263) argue: "Critical questions remain concerning the use of fMRI in lie detection. First, the pattern of activation reported in deception studies was also observed in studies of working memory, error monitoring, response selection, and 'target'detection." Wolpe et al. also draw attention to the need for scientists to outline in detail the criteria that underpin their use of the category of the "lie" in their scientific experiments. The question of the superimposition on a neurological location of a moral category such as "lie" is shown to be untenable because what is actually being imaged is a variety of physiological phenomena that cannot not effectively be localised or arrested under the banner of an extra-scientific moral category.

My framing of the "lie" as an extra-scientific moral category compels me to conclude this chapter by invoking that master interrogator of binary categories, systems of morals and positivist science, Friedrich Nietzsche. The discursive contingency and historical mutability of both truth and lies is nowhere more eloquently articulated than in Nietzsche's (1989, 250) profound meditation on the question of truth:

> What is truth? a mobile army of metaphors, metonyms, anthropomorphisms, in short, a sum of human relations which were poetically and rhetorically heightened, transferred, and adorned, and after long use seem solid, canonical, and binding to a nation. Truths are illusions about which we have forgotten that they *are* illusions, worn-out metaphors without sensory impact, coins which have lost their image and now can be used only as metal, and no longer as coins.

Nietzsche's observation on the too often effaced rhetorical dimensions of truth and lies can be effectively transposed to the conceptual infrastructure of No Lie MRI. In this neurotechnology, the digital image functions precisely as a metonym that must simultaneously disavow its metonymical status in order to function as an unmediated conduit to the truth of the digitally scanned lie. As Alan Gross (1990, 17) remarks, in science "metaphor and analogy cannot be condoned; they undercut a semantics of identity between words and things," and thereby draw attention to the constitutive role of metaphor in the scientific construction and cultural intelligibility of things. The fMRI image, as metonym, transcodes a neurological "hot spot" into a medico-legal artefact. The rhetoricity and discursivity of this move, of this ineluctable tropological turn, is what must be effaced in order for the one, the image, to be binding to the other, the neurological location of the "lie." What is operative here is what Nietzsche identified as that "capacity to volatalize visual metaphors into schema, thus to dissolve an image into a concept" (cited in Babich 1994, 97). The magical transmutation of visual metaphor (fMRI image) into schema (neurological locationism) and the assimilation of image (coloured "hot spot") into concept ("lie") bring into focus what Nietzsche (1967, 95–96) termed science's "metaphysical illusion"—"which leads science again and again to its limits at which it must turn into *art*." Precisely as physiognomics and phrenology have now been resignified as unscientific cultural artefacts, a critique of these emergent truth-detection technologies must work to disclose not only their historically and discursively contingent status, but also the manner in which the scientific logic that underpins them can effectively be seen to breach the boundaries between science and art. It is at this critical juncture that science "coils up at these boundaries and finally bites its own tail" (Nietzsche 1967, 98).

Epilogue

Biometrics' Infrastructrual Normativities and the Biopolitics of Somatic Singularities

Celebrated as offering unprecedented levels of security to contemporary culture because of their much-vaunted powers to authenticate and verify subjects, biometric technologies are expanding their reach across and into bodies in ways not previously imagined. Indeed, what is operative across the biometric industry is an exhaustive digital sectioning of the body that entails the measurement of its surfaces (hands, fingers, face, ears), its depths (retinas, irises, veins), its kinetics (gait and keystroke recognition) and its emanations (voice recognition, odor sensing and thermal recognition). In its report, *Technology Assessment: Using Biometrics for Border Security* (2002), the U.S. General Accounting Office maps a series of new and emergent biometric technologies that includes odor sensing biometrics "that can distinguish and measure body odor. This technology would use an odor-sensing instrument (an electronic 'nose') to capture the volatile chemicals that skin pores all over the body emit to make up a person's smell." The catalogue of emergent biometric technologies covers everything from ear shape recognition, nailbed identification and blood pulse biometrics to skin pattern recognition, which operates on the principle that

> skin layers differ in thickness, the interfaces between the layers have different undulations, pigmentation differs, collagen fibers and other proteins differ in density, and the capillary beds have distinct densities and locations beneath the skin. Skin pattern recognition technology measures the characteristic spectrum of an individual's skin.
>
> (U.S. General Accounting Office 2002)

As I have tracked throughout this critical analysis of biometrics, even as these emergent biometric systems signal, in their technological innovations, "the future," they simultaneously disclose their genealogical relations to previous biopolitical formations, technologies and practices. The emergent technology of ear lobe biometrics, for example, is already being touted, in typical anthropometric style, as having the capacity to reveal a screened subject's race and gender (Messaike et al. 2009). Bertillon, in the development of his "*carnet anthropométrique*" (anthropometric identity pass), had already isolated the ear as a key somatic identifier within his anthropometric system of *signalement* (description) and classification (Kaluszynski 2001, 135). A precursor form of nailbed identification was deployed, as a type of racial prophylactic, by Australian immigration officers in the early twentieth century in order to screen prospective immigrants who might "pass" as white and thus breach Australia's *Immigration Restriction Act* (1901), otherwise known as the White Australia Policy. Whenever a prospective immigrant's racial status was seen as dubious, the immigration officer would perform a racial diagnosis of the cuticles of the suspect subject's fingernails in order to detect the topical presence of melanic pigmentation. The isolation of dark pigmentation indicated the subject in question had fossilised racial deposits that, despite their seeming white appearance, atavistically marked them as having "Negroid" ancestry (Pugliese 2002, 158); they were therefore barred from entering the whites-only nation.

Similarly, both blood pulse biometrics, which measures "the blood pulse on a finger with infrared sensors," and facial thermography, in which an "Infrared camera detects heat patterns created by the branching of blood vessels and emitted from the skin" (U.S. General Accounting Office 2002), can be seen to have their nineteenth-century biopolitical prototypes. Lombroso's plethysmograph was used to record the pulse of a subject in order to ascertain whether he or she was congenitally a criminal. As Horn (2003, 128) notes, Lombroso described the plethysmograph as "'a marvelous instrument,' enabling researchers to 'descend into the penetralia of the most dissimulating man, and with an accuracy that could be called mathematical.' The plethysmograph expressed in 'millimeters,' wrote Lombroso, any emotional and psychic reaction."

Facial tomography's focus on the blood flow levels emitted by a subject's face, as diagnostic indices of their emotional, psychic and physical state, recalls Lombroso's detailed accounts of the criminological uses of blushing:

> Criminals, observed Lombroso, could not blush ... The blush, for Lombroso and other human scientists, linked the exterior of the human body with its interior, an atavistic physiology with an aberrant psychiatry ... [T]he ability or inability to blush confronted the scientist as evidence of an individual's embodied dangerousness.
>
> (Horn 2003, 107)

I juxtapose Lombroso's nineteenth-century science with a twenty-first-century account of the uses of facial thermal imaging by the U.S. Department of Defense Polygraph Institute (DoDPI):

> Thermal imaging, as a tool for credibility assessment, focuses on the use of a mid-level infrared thermal camera which looks at the spectrum of the body heat based on changes in facial blood flow. Specifically, DoDPI is seeking to determine if the changes in facial blood flow resulting from

> the activation of the sympathetic nervous system occurring when someone is anxious is because they are being deceptive.
>
> (Capps and Ryan 2009)

"Credibility assessment," DoDPI explains,

> has replaced the more narrow 'detection of deception' as it purports to encompass a much wider range of situation [sic] and contexts in which determining the existence of concealed and hidden information is vital. The continuation of these efforts and new research collaborations is leading to new and exciting avenues in the field of credibility assessment that are vital in assisting our troops in the global war on terror.
>
> (Capps and Ryan 2009)

These new biometric technologies must be genealogically situated within regimes of truth predicated on positivist ontologies of the visible. The visibility of the pulse rate and facial blood flow provide ontological evidence of criminal behaviour or intent. Mobilising these technologies in the global war on terror ensures the protection of the citizen and her or his freedom. As such, these technologies are instrumental in the biopolitical governance of subjects and in the production of freedom through the ruse of "security." "Freedom," Foucault (2008, 65) underscores, "is something which is constantly produced. Liberalism is not acceptance of freedom; it proposes to manufacture it constantly, to arouse it and produce it, with, of course, [the system] of constraints and the problems of cost raised by this production." Liberalism's investment in the manufacturing of freedom comes at the cost of producing a "political culture of danger" predicated on security: freedom is "the correlative ... of apparatuses of security" (Foucault 2008, 72 n24). Within western political cultures of danger, biometric technologies play key roles as apparatuses of security committed to the manufacturing of freedom. As technologies of truth, they verify and authenticate identities while simultaneously securing society against dangers posed by the terrorist, the criminal, the impostor, the undocumented asylum seeker, the "illegal" immigrant and so on.

In drawing attention to such emergent technologies as Project Hostile Intent, Brain Fingerprinting, No Lie MRI, blood pulse biometrics and facial tomography as technologies that promise to expose criminal intent before the fact, what can be discerned is a biopolitical investment in "divinatory" technologies of prediction, assessment of risk and pre-emption. As I have argued throughout the course of this book, despite the claims to impartiality that are made on behalf of biometric technologies, the assessment of risk is enabled by an invisibilised infrastructure predicated on regimes of normativity that, effectively, are constituted by such categories as whiteness, heteronormativity, ability, age, class and geopolitics. These *infrastructural normativities* supply the foundational categories and taxonomies that enable the biopolitical screening of subjects and the attendant "social sorting" (Lyon 2005, 13) that ensues. The freedom for some that biometric technologies produce is fundamentally predicated on the concomitant manufacture of unfreedom for targeted others. The deployment of the biometric SmartGate system by Australia and New Zealand, for example, offers the biometrically-encoded passport holders of both these countries a "simpler and more efficient [way] to cross the Tasman": "Both countries using SmartGate systems will greatly enhance the ease of use for the travelling public, which will in turn reduce processing time" (Williamson 2009). This, simultaneously, will enhance "biosecurity screening" that "will allow Customs to focus important resources on high-risk passengers" (Williamson 2009). Biometric social sorting here operates in terms of two asymmetrical categories, "trusted traveller" and "risky traveller" as identified by Louise Amoore (2006, 343; italics original):

> Whereas the *trusted traveller biometrics* tend to emphasize membership of (or inclusion in) a group based on pre-screening checks such as citizenship and past travel patterns, what I will call *immigrant biometrics* are based on ongoing surveillance and checks on patterns of behaviour. While for the trusted traveller the biometric submission is usually the end of the matter, the passport to "border lite" (if not to a borderless world), the risky traveller's biometric submission is only the beginning of a world of perennial dataveillance where the border looms large.

The biometric production of unfreedom, as the structural obverse of the freedoms promised for some by the liberal-democratic nation-state, is perhaps nowhere more clearly evidenced than at the border. The border is, after all, the very figure of demarcation and limit that biometric systems are fundamentally concerned with surveilling and controlling. Biometrics are gatekeepers of the border: non-enrolment or non-matching template literally means being precluded from crossing the threshold, the border that is being monitored and regulated by the biometric system.

Biometrics' Geocorpographies at the Border

In one of his many meditations on the complex topology of the border, Derrida (1998, 77) writes of the border as a site that is "never a secure place, it never forms an indivisible line, and it is always on the border that the most disconcerting questions get posed. Where, in fact, would a problem of topology get posed if not *on the border*? Would one ever have to worry about the border if it formed an indivisible line?" The question of topology is disconcerting because it brings into sharp focus the otherwise naturalised figure of the border—determining figure of power, placement and control. The issue of the border is one of the defining concerns of biometric technologies (van der Ploeg 2005, 115–34). As such, they are technologies preoccupied with marking, surveilling and controlling the complex topology of the threshold. Biometric systems are technologies that govern points of entry into nations, institutions, organisations, databases and so on; as such, biometric technologies must be seen as co-constitutive of the border: they are literally technologies of the border. Simultaneously, biometric systems are also constitutive of the individuating singularity of bodies: their disciplinary taxonomies, tacit knowledges and normative epistemologies (that is, their infrastructural caucacentrism, ableism, heteronormativity, classism and ageism) work to produce bodies that are either biometrically legible or not, bodies that are either precluded or enabled to cross the border. Furthermore, the topology of the border is rendered, through the application of biometric technologies, as something that is at once fixed and mobile, operating well beyond the concrete reality of the border checkpoint. The "biometric border," Amoore (2006. 338) observes, "is the portable border par excellence." Irma van der Ploeg (2005, 133) delineates the embodied dimensions of this portable biometric border:

> By virtue of the closer link established by IT-systems and biometrics between persons and their registered identities, the border becomes more than ever before part of the embodied identity of certain groups of people, verifiable at any of the many points of access to increasingly interconnected databases, and so increasingly difficult to get rid of wherever they find themselves. It thus enables the extension of the function of the border as selective and discriminating barrier beyond the actual geographical line of the inside of the country, effectively inscribed on people's bodies.

The complex biopolitical ramifications of biometric technologies that I have attempted to trace in the course of this book do not allow for facile generalisations on the power of these technologies. Whether a body is biometrically legible or not is entirely contingent on the geopolitical and somatechnical mediations of the body in question. The body that is presented for biometric enrolment is at once a *geocorpography* and a *somatechnic* entity. As a geocorpography, it signifies a body that, in any of its manifestations, is always geopolitically situated and graphically inscribed (for example, "alien," undocumented refugee, First World tourist and so on). The geopolitical significations that invest the body are constitutive of its cultural intelligibility. As a *somatechnic* entity, the body in question has always already been mediated by a series of inscriptive technologies that mark the *soma* before the fact of biometric enrolment. The inscriptive technologies of gender, ethnicity, race, sexuality, (dis)ability, class and age are not, as biometricians argue, "extrinsic" or "ancillary" to the "primary" data (for example, fingerprint or iris) of the body in question. At the moment of enrolment, the body has already been marked by an array of somatechnical mediations that have classified and categorised it in terms of its cultural intelligibility (as gendered, raced and so on) and that thereby position the subject in determinate ways at the interface of a biometric system.

That a biometric body is always already a geocorpographically and somatechnically mediated entity is effectively evidenced by the two accounts of biometric (il)legibility that I discuss later. In discussing these two different accounts, I also want to bring into focus the symbolic and physical violence that biometric technologies are implicated in producing. Across the range of biometric textbooks and manuals, biometric technologies are repeatedly represented as non-invasive, non-violent and impartial in their operations. What gets effaced in these texts is the fact that these same technologies, once imbricated within relations of biopower, are productive of different modalities of violence.

On not being biometrically legible: According to the specificity of the embodied subject in the context of their geopolitical location, everything is at stake in not being biometrically legible. I quote from Machsomwatch, an organisation of Israeli women who have painstakingly documented the plight of Palestinians at the forty checkpoints throughout the West Bank:

> The biometric examination identifies a person according to his palm print. The hands of manual labourers (in agriculture, construction, or jobs that necessitate continuous soaking of hands) are cracked and rough, and the biometric machine can't identify them. In this confrontation between sophisticated electronics and the manual labour of "hewers of wood and drawers of water," electronics have failed. The workers with their coarse and cracked hands are sent to and from the CP [checkpoint] to the DCO [district coordination officer].
>
> (Machsomwatch 2007, 45)

> We heard a soldier reprimanding a labourer and threatening that the next time he came to the checkpoint, he wouldn't be allowed through. The labourer explained to us that there's a problem with his fingerprints at the DCO—they have expired! The computer says that he is "biometrically blacklisted" and the DCO cannot understand why he continues to try to cross the checkpoint. The checkpoint commander refuses to phone the DCO to discover why the labourer has a biometric problem. The DCO representative doesn't help either. And why isn't it possible to renew fingerprints at the DCO if indeed they have "expired"?
>
> (Machsomwatch 2007, 71)

At the border, the body of the Palestinian labourer is biometrically illegible. The exertions of physical labour abrade the identificatory features of finger- and palm prints. The corporeal "valleys," "ridges" and "deltas" of the epidermis are eroded to featureless plains that render the labourer a biometric non-subject. The *soma* presented for biometric scanning cannot be geometrically processed: there are no individuating landmarks on the segment of epidermis presented to the electronic eye of the scanner. The fable of the immutable nature of corporeal signatures is belied by the permutations of the flesh, by the transformations of a body that labours, ages and sloughs off its previously enrolled versions of the self. As a result, a particular labourer's fingerprints are branded as "expired"; he is consequently "blacklisted" and shunted from bureaucratic pillar to military post. Biometric technologies can here be seen to govern and regulate the crossing of the border precisely as they constitute the biopolitical dimensions of bodies. At the border, the biometrically illegible labourer is stigmatised as a non-subject, a subaltern devoid of a "root identity." His "root identity" has been uprooted and effaced by the lapidary effects of manual labour; in the process, the Palestinian labourer is dislocated from the locus of biometric "ground truth"—it has "expired" despite the subject. Despite "inhabiting" the same body, this labourer is not identical to himself: his body is at once the same and other. Operative here is a splitting of the body from the subject: his corporeal matter no longer matches his techno-bureaucratic identity; he has become other to his biometric proxy. Incapable of supplying a corporeally legible finger or palm to the biometric system, the Palestinian labourer fails to answer the non-negotiable veridictional question—Who are you? If, in biometrics, what you are is your biometric, then this biometrically illegible Palestinian labourer is a no-body who can be sent into political drift across the indeterminate zone between the borders of nations.

On being biometrically legible: According to the specificity of the embodied subject in the context of their geopolitical location, everything is at stake in being biometrically legible. The Eurodac information system, as discussed in Chapter 3, uses fingerprint scans in order to identify, monitor and regulate the movement of asylum seekers and refugees within the European Union. Buried in a report that documents the "first coordinated inspection" of the Eurodac system is a concern regarding "cases of 'impossibility to enrol,' i.e. where the individual has no usable fingerprints" (EURODAC Supervision Coordination Group 2007, 15). The report identifies "mutilation" which is "apparently self-inflicted in more and more cases" by refugees and asylum seekers seeking to make their bodies illegible to the Eurodac biometric system (EURODAC Supervision Coordination Group 2007, 15). That the body is always already somatechnically inscribed and made legible before the fact of biometric enrolment is graphically evidenced by the refugee's compelling question: How do I render my body unreadable, non-computable and biometrically unprocessable? The refugee's traumatic answer/solution to this question brings into focus the complex interface of the *soma* and *technè*. At the border, the refugee abrades, slices and mutilates the "valleys," "ridges" and "whorls" of her fingerprints in an attempt to place her body beyond the signifying grasp of the biometric system. Through acts of self-harm, the subject's biometric "root identity" is uprooted. Through this process of self-mutilation, the subject obliterates her "ground truth." The veridictional question—"Who are you?"—is thereby rendered contingently unanswerable. As a harrowing exercise of agency, in which self-harm is deployed in order to escape identification and capture by the state's biopolitical technologies and apparatuses, it gives testimony to the despair and urgency of refugees seeking to cross the border in order to gain asylum. The disciplinary violence of these biopolitical technologies can only be evaded by the exercise of a self-inflicted violence that attempts to render the signifying attributes of the flesh unintelligible.

These acts of corporeal self-harm also bear testimony to something that appears to escape techno-rationalist conceptualisations of the body. On the one hand, the body is always already somatechnologised through a series of cultural and inscriptive mediations. These somatechnical mediations establish corporeal atlases, typologies and topographies. The entirety of the body in question is mapped out in the techno-rationalist light of these discourses. So, for example, the identificatory attributes of the finger are, in finger-scan biometrics, described in terms of typologies (loop, delta and whorl) and topographies (ridges and valleys) that establish its computational, geometric and algorithmic conditions of signification. Yet, simultaneously, something other remains, a corporeal residue that is recalcitrant to all these mediations. The cut of the blade across the flesh is a somatechnical act that enables the reconfiguring of the epidermis into that obtuse figure of the scar, disordered knitting of skin that obliterates its own typologies and topographies. The refugee's act of self-mutilation mobilises this asignifying residue of flesh in order to fold the body upon itself and produce a moment of scarring occlusion and contingent non-meaning.

In these contexts, biometric technologies function as disciplinary apparatuses fundamental to the biopolitical operations of the nation-state. "Disciplinary power," Foucault (2006, 55) argues, "is individualizing because it fastens the subject-function to the somatic singularity by means of … a system of pangraphic panopticism." Pangraphic panopticism articulates the digital corpo-graphing of biometric signatures into interoperable biometric databases networked across nation-states and multiple institutions. As biopolitical instantiations of disciplinary power, biometric systems are instrumental in fastening subject-functions to the somatic singularity of the persons in question: for example, Palestinian labourer or undocumented refugee. "[I]t is insofar," Foucault (2006, 56) writes, "as the somatic singularity became the bearer of the subject-function through disciplinary mechanisms that the individual appeared within a political system." In the two contexts of biometric (il)legibility that I have discussed, the disciplinary mechanisms of biometrics proceed to fasten to the somatic singularity of the Palestinian labourer and the undocumented refugee the following biopolitically determined subject-functions: *non-person* (devoid of rights) and *illegal person* (transgressing the borders of the nation-state).

The two stories that I have recounted emerge as perverse mirror-images of each other, structured by inversions that are marked by (il)legible differences that yet betray uncanny homologies. I juxtapose these two stories in the concluding gestures of this book in order to begin to unsettle celebratory, technophile accounts of biometrics by marking those residual, marginal, recalcitrant, non-normative bodies that would otherwise not signify. Those bodies, for example, that are branded with the stigma of "disabled," that are consequently dismissed as not worthy of being biometrically accounted for and are, literally, excised from the signifying corpus of biometric enrolment rates, even as they create minor and transitory "blockages" within the everyday operations of biometric systems. These biometrically illegible subjects, and their non-normative bodies, include a certain Mr S, a cancer patient taking the common cancer drug capecitabine. One of the side effects of capecitabine is that it obliterates a person's fingerprints because it causes redness and peeling of the skin. Arriving in the U.S. from Singapore, Mr S was held for four hours by U.S. customs officials "before deciding he was not a security threat" (Cheng 2009, 3). In this instance, the pathology of the flesh and a prescribed pharmacological regime produce a non-normative subject devoid of fingerprints: they have been chemically exfoliated, leaving no identificatory epidermal landmarks. The infrastructural normativity of the biometric system, when confronted by such a non-normative subject, automatically brands him a "security threat," and he is thereby detained for four hours.

Biometrics' Infrastructural Normativities and the Biopolitics of Somatic Singularities

Invited in March 2004, to deliver a series of guest lectures at New York University, the influential Italian philosopher, Giorgio Agamben, declined to go because he objected to having his fingerprints biometrically processed in order to gain entry into the U. S. Soon after declining this invitation, he wrote a short piece explaining his decision. Drawing attention to "Today's electronically enhanced possibilities of the state to exercise control over its citizens," Agamben (2004) writes that "there is one threshold in the control and manipulation of bodies, the transgression of which would signify a new global condition ... The electronic registration of finger prints, the subcutaneous tattoo and other such practices must be located on that threshold." In the face of these multiple practices of "electronic registration," Agamben (2004) concludes that: "What we are witnessing is no longer the free and active participation on the political level, but the appropriation and registration of the most private and unsheltered element, that is the biological life of bodies." While I acknowledge the symbolic importance of his stand in refusing to be biometrically scanned, I am surprised by Agamben's perception that this biometric registration of bodies is precluding an otherwise "free and active participation on the political level," as if prior to the contemporary deployment of biometrics all subjects were equally empowered—the histories of colonialism, racism, sexism, homophobia, ableism and classism show otherwise.

As I have attempted to demonstrate in the course of this book, contemporary biometric systems are the culmination of a series of anthropometric technologies that can be genealogically traced back to the early nineteenth century and the historical emergence of biopolitics. Once situated in this context, contemporary biometric "appropriation and registration of the most private and unsheltered element, that is the biological life of bodies," must be seen as having been in process for well over one hundred-and-fifty years, beginning with the appropriation and registration of the fingerprints of members of various Indian "tribes" by the British in colonial India, and continuing, in that boomerang effect that I discussed in Chapter 1, through the export of this biometric technology back to metropolitan Europe. The subsequent development of Bertillon's police anthropometry firmly placed the biological life of target subjects within indexed filing systems that proved to be foundational to the operations of the biopolitical state. Bertillon's biometric system, predicated on an exhaustive registration of the biological life of suspect bodies, enabled the classification, surveillance and punishment of Europe's internal others. Evidenced here is a type of genealogical continuity between past and contemporary biopolitical technologies, apparatuses and disciplinary practices that seems to escape Agamben. If there is a genealogical rupture with past biopolitical systems deployed by the state and the crossing of a new threshold, it is evidenced by the development of networked databases of biometric interoperability that now function to constitute mobile, transnational and transinstitutional flows of the biometrically processed "biological life of bodies."

In his concluding observations on the spread of biometric systems in contemporary societies, Agamben (2004) notes that:

> Paradoxically, the citizen is thus rendered a suspect all along, a suspect against which all those techniques and installations need to be mounted that had originally been conceived of only for the most dangerous individuals. *Per definition*, mankind has been declared the most dangerous of all classes.

Agamben's concluding observation effectively obliterates difference and homogenises the specificities of somatechnically mediated and geocorpographically positioned subjects. Within the increasingly networked operations of contemporary biometric technologies, not all citizens are rendered suspect and dangerous, and thus not all citizens are equally caught within biopolitical systems of biometric surveillance, discrimination and disenfranchisement. To a degree, this rendering of *all* citizens as suspect and dangerous is tantamount to a disavowal of the very categories of whiteness, heteronormativity, ableism, class and geopolitics that, in practice, continue to inform the normative infrastructures of contemporary biometric technologies and that enable practices of enfranchisement for some and disenfranchisement for others.

The animating question of biopolitics—"Who are you?"—is, as I have demonstrated, the foundational question of all biometric technologies. What remains unsaid, in the biometric processing of this question, is that who you are is often made to be coextensive with *what you are*: a Palestinian manual labourer, a transgender subject, a person with a disability, an undocumented refugee from the Global South or a person of colour. These are not discrete categories as, in practice, they so often intersect and inscribe the one body. At the point of biometric enrolment, the body in question is at once a somatic singularity and a somatechnically mediated and geocorpographically positioned figure enmeshed within normative and disciplinary networks of biopolitical power.

References

Agamben, Giorgio. 1998. *Homo Sacer: Sovereign Power and Bare Life*. Stanford: Stanford University Press.

——2004. Bodies without Words: Against the Biopolitical Tattoo. *German Law Journal* 5. http://www.germanlawjournal.de/print.php?id=371.

Aldridge, Susan. 1996. *The Thread of Life: The Story of Genes and Genetic Engineering*. Cambridge: Cambridge University Press.

Alloway, Kevin D. and Thomas C. Pritchard. 2007. *Medical Neuroscience*. Raleigh, N.C.: Hayes Barton Press.

Amoore, Louise. 2006. Biometric Borders: Governing Mobilities in the War on Terror. *Political Geography* 25: 336–51.

Arogundade, Ben. 2000. *Black Beauty*. London: Pavillion.

Associated Press. 2002. Irises, voices give away terrorists. http://www.cnn.com/2002/TECH/ptech/11/07/terror.biometrics.ap/index.html.

Babich, Babette E. 1994. *Nietzsche's Philosophy of Science*. Albany: State University of New York Press.

Banton, Michael. 1987. *Racial Theories*. Cambridge: Cambridge University Press.

Barker, Francis. 1984. *The Tremulous Private Body*. London: Methuen.

Barthes, Roland. 1993. *Camera Lucida*. London: Vintage.

BBC. 2008. Italy fingerprint plan criticised. BBC News, 26 June 2008. http://news.bbc.co.uk/2/hi/europe/7476413.stm.

Beauchamp, Toby. 2009. Artful Concealment and Strategic Visibility: Transgender Bodies and U.S. State Surveillance after 9/11. *Surveillance and Society* 6: 356–66.

Beavan, Colin. 2001. *Fingerprints: The Origins of Crime Detection and the Murder Case That Launched Forensic Science*. New York: Hyperion.

Bernal, Martin. 1987. *Black Athena*. London: Free Association Books.

Bianchini, Marcello Levi. 1906. La mentalità della razza calabrese: saggio di psicologia etnica. *Rivista Psicologica Applicata alla Pedagogia ed alla Psicopatologia* 2: 13–21.

Biber, Katherine. 2005. Your Fantasy, My Crime: Aboriginality, Crime and Photographs. Unrequited Justice.http://www.ccs.mq.edu.au/justice/papers.html.

Bindman, David. 2002. *From Ape to Apollo: Aesthetics and the Idea of Race in the 18th Century*. London: Reaktion Books.

Brain Fingerprinting Laboratories.

http://www.brainwavescience.com/TechnologyOverview.php.

Burke, Anthony. 2001. *In Fear of Security*. Annandale, NSW: Pluto Press.

Camper, Petrus. 1794 [1791]. *A Treatise on the Natural Difference of Features in Persons of Different Countries and Periods of Life; and on Beauty, as Exhibited in*

Ancient Sculpture: With a New Method of Sketching Heads, National Features, and Portraits of Individuals, with Accuracy. London: C. Dilly.

Caplan, Jane and John Torpey. 2001. Introduction. In *Documenting Individual Identity*, eds. Jane Caplan and John Torpey, 1–12. Princeton and Oxford: Princeton University Press.

Caplan, Jane. 2001. "This or That Particular Person": Protocols of Identification in Nineteenth-Century Europe. In *Documenting Individual Identity*, eds. Jane Caplan and John Torpey, 49–66. Princeton and Oxford: Princeton University Press.

Capps, John G. and Andrew Ryan. 2009. It's Not Just a Polygraph Anymore. APA Online: Psychological Science Agenda. http://www.apa.org/science/psa/polygraph_print.html.

Cartwright, Lisa. 1995. *Screening the Body: Tracing Medicine's Visual Culture*. Minneapolis: University of Minnesota Press.

Cephos Corp. http://www.cephoscorp.com/index.html.

Cheng, Maria. 2009. Drug Erases Fingerprints, Causing Immigration Drama. *Sydney Morning Herald*, 28 May: 3.

Choi, Alex Hwansoo and Chung Nguyen Tran. 2008. Hand Vascular Pattern Technology. In *Handbook of Biometrics*, ed. Anil K. Jain, Patrick Flynn and Arun A. Ross, 253–70. New York: Springer.

Cole, Simon A. 2002. *Suspect Identities: A History of Fingerprinting and Criminal Identifications*. Cambridge, MA and London: Harvard University Press.

Coleman, William. 1971. *Biology in the Nineteenth Century*. Cambridge: Cambridge University Press.

Combe, George, and Johann Gaspar Spurzheim. 2007 [1834]. *Readings in Phrenology*. Ed. Frederick H. Hurd. n. p.: Sierra Madre Bookshop.

Combe, George. 2008 [1834]. *Elements of Phrenology*. Ed. Fredrick H. Hurd. n. p.: Sierra Madre Bookshop.

Compact Oxford English Dictionary. 1992. Oxford: Clarendon Press.

Council of Australian Governments' Communiqué: Special Meeting on Counter-Terrorism. 2005. http://parlinfo.aph.gov.au:80/parlInfo/download/media/pressrel/E4FH6/upload_binary/e4fh61.pdf;fileType%3Dapplication%2Fpdf.

Crary, Jonathan. 1998. *Techniques of the Observer*. Cambridge, MA and London: MIT Press.

Critchley, Simon. 1992. *The Ethics of Deconstruction: Derrida and Levinas*. Oxford: Blackwell.

Crompton, Malcolm. 2002. Biometrics and Privacy. *Privacy Law and Policy Reporter* 32. http://search.austlii.edu.au/au/journals/PLPR/2002/32/html.

Cunneen, Chris. 2005. Colonialism and Historical Injustice: Reparations for Indigenous Peoples. *Social Semiotics* 15: 59–80.

Davis, Angela Y. 1998. The *Angela Y. Davis Reader*. Ed. Joy James. Malden: Blackwell.

Davis, F. James. 1991. *Who Is Black?* University Park, Penn.: University of Pennsylvania Press.

Davis, Lennard A. 2006. Constructing Normalcy. In *The Disability Studies Reader*, ed. Lennard A. Davis, 3–16. New York and London: Routledge.

De Certeau, Michel. 1988. *The Practice of Everyday Life*. Berkeley: University of California Press.

De Lauretis, Teresa. 1987. *Technologies of Gender*. Bloomington: Indiana University Press.

De Waal Malefijt, Annemarie. 1974. *Images of Man*. New York: Alfred A. Knopff.
Department of Immigration and Multicultural and Indigenous Affairs [DIMIA]. 2004. Fact Sheet No. 84 "Biometric Initiatives." http://www.immi.gov.au/facts/84biometric.htm.
Derrida, Jacques. 1976. *Of Grammatology*. Baltimore and London: Johns Hopkins University Press.
——1986. *Margins of Philosophy*. Brighton: Harvester Press.
——1990. *Limited Inc*. Evanston: Northwestern University Press.
——1992. How to Avoid Speaking: Denials. In *Derrida and Negative Theology*, ed. Harold Coward and Toby Foshay, 73–142. Albany: State University of New York Press.
——1998. *Resistances of Psychoanalysis*. Stanford: Stanford University Press.
——2002a. *Negotiations*. Stanford: Stanford University Press.
——2002b. *Without Alibi*. Stanford: Stanford University Press.
Derrida, Jacques and Bernard Stiegler. 2002. *Echographies of Television*. Cambridge: Polity Press.
Dreyfus, Hubert L. and Paul Rabinow. 1982. *Michel Foucault: Beyond Structuralism and Hermeneutics*. New York: Harvester Wheatsheaf.
Dyer, Richard. 1997. *White*. New York and London: Routledge.
Eisenberg, Carol. 2007. Homeland Security Exploring Mass-Screening System. Newsday.com, 15 August. http://www.newsday.com/news/nationworld/ny-usterr15332662aug15,0,7332097.story.
Elkins, James. 1999. *Pictures of the Body*. Stanford: Stanford University Press.
Ellison, Ralph. 1972. *Invisible Man*. New York: Random House.
EURODAC Supervision Coordination Group. 2007. Report of the First Coordinated Inspection. www.dsk.gv.at/DocView.axd?CobId=35790.
Fanon, Frantz. 1970[1952]. *Black Skin, White Masks*. London: Paladin.
——2004. *The Wretched of the Earth*. New York: Grove Press.
Feng, Li, Jianhuang Lai and Lei Zhang. 2004. 3D Surface Reconstruction Based on One Non-Symmetric Face Image. In *Advances in Biometric Person Authentication: 5th Chinese Conference on Biometric Recognition, SINOBIOMETRICS 2004, Guangzhou, China, December 2004, Proceedings*, ed. Stan Z. Li, Jianhuang Lai, Tieniu Tan, Guocan Feng and Yunhong Wang, 268–74. Berlin, Heidelberg, New York: Springer.
Foucault, Michel. 1973. *The Order of Things: An Archaeology of the Human Sciences*. New York: Vintage Books.
——1975. *The Birth of the Clinic: An Archaeology of Medical Perception*. New York: Vintage.
——1980. *Power/Knowledge*. Ed. Colin Gordon. New York: Pantheon Books.
——1982. *Discipline and Punish: The Birth of the Prison*. Harmondsworth: Penguin.
——1985. *The Archaeology of Knowledge*. London: Tavistock Publications.
——1986. *Language, Counter-Memory, Practice*. Ed. D. F. Bouchard. Ithaca, NY: Cornell University Press.
——1990. *The History of Sexuality, Vol. 1*. London: Penguin.
——2003. *"Society Must Be Defended": Lectures at the Collège de France, 1975–1976*. New York: Picador.
——2003a. *Abnormal: Lectures at the Collège de France 1974–1975*. New York: Picador.
——2006. *Psychiatric Power: Lectures at the Collège de France 1973–1974*. Basingstoke and New York: Palgrave Macmillan.

——2007. *Security, Territory, Population: Lectures at the Collège de France 1977–1978*. Basingstoke and New York: Palgrave Macmillan.

——2008. *The Birth of Biopolitics: Lectures at the Collège de France, 1978–79*. Basingstoke and New York: Palgrave Macmillan.

Frankenberg, Ruth. 1993. *The Social Construction of Whiteness*. Minneapolis and London: Routledge.

Fusco, Coco. 2001. *The Bodies That Were Not Ours*. London and New York: Routledge.

Gambino, R. 1974. *Blood of My Blood*. New York: Doubleday.

Garland, Brent. 2004. Neuroscience and the Law: A Report. In *Neuroscience and the Law*, ed. Brent Garland, 1–47. New York: Dana Press.

Gates, Kelly A. 2005. Biometrics and Post-9/11 Technologies. *Social Text* 83: 35–53.

Gates, Kelly. 2006. Identifying the 9/11 'Faces of Terror': The Promise and Problem of Facial Recognition Technology. *Cultural Studies* 20: 417–40.

Geracimos, Ann. 2004. Walking "Signature." Washington Times Online, March 4. http://www.cat.nyu.edu/current/news/media/Wash_times.htm.

Geyl, Pieter. 1968. *The Netherlands in the Seventeenth Century*. London: Ernest Benn.

Gibson, Mary. 2002. *Born to Crime: Cesare Lombroso and the Origins of Biological Criminology*. Westport: Praeger.

Gilbert, Kevin. 1988. Introduction. In *Inside Black Australia*, ed. Kevin Gilbert, xv–xxiv. Ringwood, Vic.: Penguin.

Gilman, Sander L. 1986. Black Bodies, White Bodies: Towards an Iconography of Female Sexuality in Late Nineteenth-Century, Art, Medicine, and Literature. In *"Race," Writing, and Identity*, ed. Henry Louis Gates, Jr., 223–61. Chicago: University of Chicago Press.

Gilroy, Paul. 2004. *Between Camps: Nations, Cultures and the Allure of Race*. London and New York: Routledge.

Goodrich, Peter. 1990. *Languages of Law*. London: Weidenfeld and Nicolson.

Gordon, Evian, Chris Rennie, Arthur Toga and John Mazziotta. 2000. Human Brain Imaging Technologies. In *Integrative Neuroscience: Bringing Together Biological, Psychological, and Clinical Models of the Brain*, ed. Evian Gordon, 233–44. Amsterdam: Harwood Academic Publishers.

Gould, Stephen Jay. 1987. Petrus Camper's Angle. *Natural History* 96: 12–18.

——1996. *The Mismeasure of Man*. New York: W.W. Norton and Company.

Greely, Henry T. and Judy Illes. 2007. Neuroscience-Based Lie Detection: The Urgent Need for Regulation. *American Journal of Law and Medicine* 33: 377–421.

Groebner, Valentin. 2007. *Who Are You? Identification, Deception, and Surveillance in Early Modern Europe*. New York: Zone Books.

Gross, Alan N. 1990. *The Rhetoric of Science*. Cambridge, MA and London: Harvard University Press.

Guglielmo, J. and S. Salerno. 2003. *Are Italians White?* New York and London: Routledge.

Hall, Stuart. 1993. The Question of Cultural Identity. In *Modernity and Its Futures*, ed. S. Hall, D. Held and T. McGrew, 273–326. Trowbridge: Polity Press and the Open University.

Haraway, Donna. 1991. *Simians, Cyborgs, and Women: The Reinvention of Nature*. New York: Routledge.

Harris, David. 2002. *Profiles in Injustice*. New York: The New Press.

Hartley, Lucy. 2001. *Physiognomy and the Meaning of Expression in Nineteenth-Century Culture*. Cambridge: Cambridge University Press.

Hayles, Katherine N. 1999. *How We Became Posthuman*. Chicago: Chicago University Press.

Holbert, Steve and Lisa Rose. 2004. *The Color of Guilt and Innocence: Racial Profiling and Police Practices in America*. San Ramon: Page Marque Press.

hooks, bell. 1992. *Black Looks: Race and Representation*. Boston: South End Press.

Horn, David. 2003. *The Criminal Body: Lombroso and the Anatomy of Deviance*. New York and London: Routledge.

Hrdlicka, Ales. 1939. *Practical Anthropometry*. Philadelphia: The Wistar Institute of Anatomy and Biology.

International Association of Chiefs of Police (IACP). 2005. Training Keys #581: Suicide (Homicide) Bombers: Part 1. www.theiacp.org/pubinfo/IACP581Suicide-BombersPart1.pdf.

Jackson, John P. Jr., and Nadine M. Weidman. 2006. *Race, Racism and Science*. New Brunswick, NJ: Rutgers University Press.

Jacobson, M. P. 1998. *Whiteness of a Different Colour*. Cambridge, MA: Harvard University Press.

Jain, Anil K. and Arun Ross. 2004. Multibiometric Systems. *Communications of the ACM* 47: 34–40.

——2008. Introduction to Biometrics. In *Handbook of Biometrics*, ed. Anil K. Jain, Patrick Flynn and Arun A. Ross, 1–22. New York: Springer.

Jamail, Dahr. 2007. *Beyond the Green Zone: Dispatches from an Unembedded Journalist in Occupied Iraq*. Chicago: Haymarket Books.

Jordanova, Ludmilla. 1993. The Art and Science of Seeing Medicine: Physiognomy 1780–1820. In *Medicine and the Five Senses*, ed. W. F. Bynum and R. Porter, 122–33. Cambridge: Cambridge University Press.

Jupp, James. 1991. *Immigration*. Sydney: Sydney University Press.

Kaluszynski, Martine. 2001. Republican Identity: Bertillonage as Government Technique. In *Documenting Individual Identity*, ed. Jane Caplan and John Torpey, 123–38. Princeton: Princeton University Press.

Kari, Lila and Laura F. Landweber. 2000. Computing with DNA. In *Bioinformatics: Methods and Protocols*, ed. Stephen Misener and Stephen A. Krawetz, 413–30. New Jersey: Humana Press.

Kennedy, Donald. 2005. Neuroimaging: Revolutionary Research Tool or a Post-Modern Phrenology? *American Journal of Bioethics* 5: 19.

Kevles, Daniel J. 2004. *In the Name of Eugenics*. Cambridge, MA: Harvard University Press.

Kevorkian, Martin. 2006. *Color Monitors: The Black Face of Technology in America*. Ithaca: Cornell University Press.

Knight, Bernard. 1997. *Simpson's Forensic Medicine*. London: Arnold.

Kress, Gunther and Theo van Leeuwen. 2002. Colour As a Semiotic Mode: Notes for a Grammar of Colour. *Visual Communication* 1: 343–68.

Lalvani, Suren. 1996. *Photography, Vision, and the Production of Modern Bodies*. Albany, NY: State University of New York Press.

Langleben, Daniel D., James W. Loughead, Warren B. Bilker, Kosha Ruparel, Anna Rose Childress, Samantha I. Busch, and Ruben C. Gur. 2005. Telling Truth from Lie in Individual Subjects with Fast Even-Related fMRI. *Human Brain Mapping* 26: 262–72.

Lao, Shihong and Masato Kawade. 2004. Vision-Based Face Understanding Technologies and Their Application. In *Advances in Biometric Person Authentication: 5th Chinese Conference on Biometric Recognition, SINOBIOMETRICS 2004, Guangzhou, China, December 2004, Proceedings*, ed. Stan Z. Li, Jianhuang Lai, Tieniu Tan, Guocan Feng and Yunhong Wang, 339–48. Berlin, Heidelberg, New York: Springer.

Levinas, Emmanuel. 1988. *Totality and Infinity*. Pittsburgh: Duquesne University Press.

——1991. *Otherwise Than Being or Beyond Essence*. Dordrecht: Kluwer Academic Publishers.

——1998. *Discovering Existence with Husserl*. Evanston: Northwestern University Press.

——2003. *On Escape/De l'évasion*. Stanford: Stanford University Press.

Li, Stan Z., Ben Schouten and Massimo Tistarelli. 2009. Biometrics at a Distance: Issues, Challenges, and Prospects. In *Handbook of Remote Biometrics for Surveillance and Security*, ed. Massimo Tistarelli, Stan Z. Li and Rama Chellappa, 3–21. Dordrecht: Springer.

Lombroso, Cesare and Guglielmo Ferrero. 2004 [1893]. *Criminal Woman, the Prostitute, and the Normal Woman*. Trans. N. H. Rafter and M. Gibson. Durham: Duke University Press.

Lombroso, Cesare. 1889. *L'Uomo Delinquente in Rapporto all'Antropologia, alla Giurisprudenza ed alle Discipline Carcerarie*, 2 volumes. Torino: Fratelli Bocca.

——2007 [1876–97]. *Criminal Man*. Trans. M. Gibson and N. H. Rafter. Durham: Duke University Press.

López, Ian F. Haney. 1996. *White by Law*. New York: New York University Press.

Lyon, David. 2003. *Surveillance After September 11*. Cambridge: Polity.

——2004. *The Electronic Eye: The Rise of Surveillance Society*. Minneapolis: University of Minnesota Press.

——2005. Surveillance as Social Sorting: Computer Codes and Mobile Bodies. In *Surveillance as Social Sorting*, ed. David Lyon, 13–30. Abingdon, Oxon and New York: Routledge.

——2008. *Surveillance Studies: An Overview*. Cambridge: Polity.

——2008a. Biometrics, Identification and Surveillance. *Bioethics* 22: 499–508.

Machsomwatch. 2007. *Machsomwatch Alerts 2007: Yearly Report*. Jerusalem: Machsomwatch.

Mbembe, Achille. 2003. Necropolitics. *Public Culture* 15: 11–40.

McDonough Presbyterian Church. 2003. They Will Know We Are Christians by Our Walk. http://www.mcdonoughpresbyterian.com/Sermons/they_will_know_we_are_christians_by_our_walk.htm.

Mercer, Kobena. 1990. Black Hair/Style Politics. In *Out There: Marginalization and Contemporary Cultures*, ed. R. Ferguson, M. Gever, T. T. Minh-ha and C. West, 247–64. New York and Cambridge, MA: New Museum of Contemporary Art and MIT Press.

——1994. *Welcome to the Jungle: New Positions in Black Cultural Studies*. New York and London: Routledge.

Messaike, Elias, Meiya Sutsino, Fraser Torpy and Tamara Szlynda. 2009. *Mapping the Human Auricle in Asians Residing in the Sydney Region*. Department of Cell and Molecular Biology, University of Technology, Sydney, Skullforensics, N.S.W. http://docs.google.com/viewer?a=v&q=cache:LEFQRVtNnRwJ:195.95.2.80/

cps_new/index.php%3Foption%3Dcom_docman%26task%3Ddoc_download%26gid%3D109+Ear+biometrics+mapping+the+human+auricle+in+asians+in+residing&hl=en&pid=bl&srcid=ADGEEShVpm-O-LQay9k1eglBLghB3FLsO4l5OnMPNEnAx-EQTAS9mujViwyuFXf7zHC2kZmvew3oJVhOm6ty-loWrN9vAF0mub8XXMJNTe7QGSXOcAu5UmXJT_c-iIvKNKmirtegTqHkh3&sig = AHIEtbSzzfByYn9Celw8jLut7bSOIWtMlA.

Mitchell, William J. 1998. *The Reconfigured Eye: Visual Truth in the Post-Photographic Era*. Cambridge, MA and London: MIT Press.

Moreno, Jonathan D. 2006. *Mind Wars: Brain Research and National Defense*. New York and Washington, DC: Dana Press.

Moriarty, Jane Campbell, J. D.,. 2008. Flickering Admissibility: Neuroimaging Evidence in the U.S. Courts. *Behavioural Science and the Law* 26: 29–49.

Morrison, Toni. 1992. *Playing in the Dark: Whiteness and the Literary Imagination*. London: Picador.

Moses, Dirk A. 2005. Genocide and Settler Society in Australian History. In *Genocide and Settler Society*, ed. Dirk A. Moses, 3–48. New York and Oxford: Berghahn Books.

Mosse, George. 1978. *Toward the Final Solution: A History of European Racism*. London: Dent.

Mulvey, Laura. 1999 [1975]. Visual Pleasure and Narrative Cinema. In *Visual Culture: The Reader*, ed. Jessica Evans and Stuart Hall, 381–89. London: Sage.

Murphy, Marina. 2004. Infallible Witness. Brain Fingerprinting Laboratories. http://www.brainwavesciences.com/.

Nanavati, Samir, Michael Thieme and Raj Nanavati. 2002. *Biometrics: Identity Verification in a Networked World*. New York: John Wiley and Sons.

Nandakumar, Karthik, Arun Ross and Anil K. Jain. 2008. Incorporating Ancillary Information in Multibiometric Systems. In *Handbook of Biometrics*, ed. Anil K. Jain, Patrick Flynn and Arun A. Ross, 335–55. New York: Springer.

National Science and Security Council. 2006. Vascular Pattern Recognition. http://www.biometricscatalog.org/NSTCSubcommittee.

National Science and Technology Council.(NSTC) 2006. Biometrics History. www.biometrics.gov/Documents/BioHistory.pdf.

Negra, Diane. 2001. *Off-White Hollywood*. London and New York: Routledge.

Neville, Alisoun. 2006. Classification, Denial and the Racial State: *Cubillo v Commonwealth*. PhD dissertation, La Trobe University, Melbourne.

New York Times. 2007. Trying to Distinguish Friend From Foe on Baghdad's Streets. April 4 April, A7.

NewsCenter. 2007. Technology Would Help Detect Terrorists before They Strike. The State University of New York at Buffalo. http://www.buffalo.edu/news/8879.

Nietzsche, Friedrich. 1966. *Beyond Good and Evil*. New York: Vintage.

——1967. *The Birth of Tragedy and The Case of Wagner*. New York: Vintage Books.

——1989. *Friedrich Nietzsche on Rhetoric and Language*. Ed. Sander L. Gilman, Carole Blair and David J. Parent. New York and Oxford: Oxford University Press.

Nilsen, Alf Gunvald. 2008. Email message to author, 27 June.

Nixon, Kristin Adair, Valerio Aimale and Robert K. Rowe. 2008. Spoof Detection Schemes. In *Handbook of Biometrics*, ed. Anil K. Jain, Patrick Flynn and Arun A. Ross, 403–23. New York: Springer.

No Lie MRI. http://www.noliemri.com/.

Office of the Under Secretary of Defense for Acquisition, Technology, and Logistics. 2007. Report of the Defense Science Board Task Force on Defense Biometrics. Washington, DC: U.S. Department of Defense.

Osuri, Goldie. 2009. Necropolitical Complicities: (Re)Constructing a Normative Somatechnics of Iraq. *Social Semiotics* 19: 31–45.

Palmer, Alison. 2000. *Colonial Genocide*. Adelaide: Crawford House Publishing.

Parenti, Christian. 2003. *The Soft Cage: Surveillance in America from Slavery to the War on Terror*. New York: Basic Books.

Parziale, Giuseppe and Yi Chen. 2009. Advanced Technologies for Touchless Fingerprint Recognition. In *Handbook of Remote Biometrics for Surveillance and Security*, ed. Massimo Tistarelli, Stan Z. Li and Rama Chellappa, 83–109. Dordrecht: Springer.

Pearson, Karl. 1901. *National Life from the Standpoint of Science*. London: Adam and Charles Black.

Perera, Suvendrini. 1993. Representation Wars: Malaysia, *Embassy*, and Australia's *Corps Diplomatique*. In *Australian Cultural Studies*, ed. J. Frow and M. Morris, 15–29. Sydney: Allen and Unwin.

——. 2002. What Is a Camp ... ? *Borderlands* 1. http://www.borderlandsejournal.adeliade.edu.au/vol1no1_2002/perera_camp.html.

——2002a. A Line in the Sea: the *Tampa*, Boat Stories and the Border. *Cultural Studies Review* 8: 11–27.

——2006. "They Give Evidence": Bodies, Borders and the Disappeared. *Social Identities* 12: 637–56.

——2008. A Pacific Zone? (In)Security, Sovereignty, and Stories of the Pacific Borderscape. In *Borderscapes: Hidden Geographies and Politics at Territory's Edge*, ed. P. K. Rajaram and C. Grundy-Warr, 201–27. Minneapolis: University of Minnesota Press.

——2009. *Australia and the Insular Imagination: Beaches, Borders, Boats, and Bodies*. New York: Palgrave Macmillan.

Potts, Alex. 2000. *Flesh and the Ideal: Winckelmann and the Origins of Art History*. New Haven: Yale University Press.

Prosser, Jay. 1998. *Second Skins: The Body Narratives of Transsexuality*. New York: Columbia University Press.

Pugliese, Joseph. 1999. Identity in Question: A Grammatology of DNA and Forensic Genetics. *International Journal for the Semiotics of Law/Revue International de Sémiotique Juridique* 12: 419–44.

——2002. Race As Category Crisis: Whiteness and the Topical Assignation of Race. *Social Semiotics* 12: 149–68.

——2003. The Locus of the Non: The Racial Fault-Line "of Middle Eastern Appearance." *Borderlands* 2, http://www.borderlandsejournal.adelaide.edu.au/vol2no3_2003/pugliese.html.

——2005. "Demonstrative Evidence": A Genealogy of the Racial Iconography of Forensic Art and Illustration. *Law and Critique* 15: 283–320.

——2005a. Necrological Whiteness: The Racial Prosthetics of Template Bodies. *Continuum* 19: 349–64.

——2007. Geocorpographies of Torture. *Australian Critical Race and Whiteness Studies Association Journal* 3, http://www.acrawsa.org.au/.

Pugliese, Joseph and Susan Stryker. 2009. Introduction: The Somatechnics of Race and Whiteness. *Social Semiotics* 19: 1–8.

Report of the National Inquiry into the Separation of Aboriginal and Torres Strait Islander Children from Their Families. 1997. *Bringing Them Home*. Sydney: Sterling Press.

Rivers, Christopher. 1994. *Face Value: Physiognomical Thought and the Legible Body in Marivaux, Lavater, Balzac, Gautier, and Zola*. Madison: University of Wisconsin Press.

Roddy, A. R. and J. D. Stosz. 1999. Fingerprint Feature Processing Techniques and Poroscopy. In *Intelligent Biometric Techniques in Fingerprint and Face Recognition*, ed. L. C. Jain, U. Halici, I. Hayashi, S. B. Lee and S. Tsutsui, 35–105. Boca Raton and London: CRC.

Roediger, David. 1994. *Towards the Abolition of Whiteness*. London: Verso.

Rolland, Jacques. 2003. Getting Out of Being by a New Path. In *On Escape/De l'évasion*, Emmanuel Levinas, 3–48. Stanford: Stanford University Press.

Rony, Fatimah Tobing. 1998. *The Third Eye: Race, Cinema, and the Ethnographic Spectacle*. Durham and London: Duke University Press.

Ross, Arun and Norman Poh. 2009. Multibiometric Systems: Overview, Case Studies, and Open Issues. In *Handbook of Remote Biometrics for Surveillance and Security*, ed. Massimo Tistarelli, Stan Z. Li and Rama Chellappa, 273–92. Dordrecht: Springer.

Rothstein, Linda. 2003. Something in the Way She Walked? *Bulletin of the Atomic Scientists*. January 1. www.highbeam.com/doc/1G1–96268095.html.

Russell, Lynette. 2001. *Savage Imaginings: Historical and Contemporary Constructions of Australian Aboriginalities*. Melbourne: Australian Scholarly Publishing.

Said, Edward. 1991. *Orientalism*. Harmondsworth: Penguin.

——1992. *The Question of Palestine*. New York: Vintage.

Sample, Ian. 2007. Security Firms Working on Devices to Spot Would-Be Terrorists in Crowd. *The Guardian*, 9 August. http://global.factiva.com/ha/default.aspx.

Sappol, Michael. 2002. *A Traffic of Dead Bodies*. Princeton, NJ: Princeton University Press.

Schwartz-Dupre, Rae Lynn. 2007. Rhetorically Representing Public Policy. *Feminist Media Studies* 7: 433–53.

Seele, Peter. 2007. Is Blue the New Green? Colors of the Earth in Corporate PR and Advertisement to Communicate Ethical Commitments and Responsibility. Working Papers of the Center for Responsibility Research, Jahrgang 1/2007, Heft 3. http://www.responsibility-research.de/CRRFForschung.html.

Sejoe, Mmaskepe M. 2009. Conference discussion at the Human Rights Law and Practice Conference, 20–21 October, Melbourne, Australia.

Shorter Oxford English Dictionary. 1978. Oxford: Clarendon Press.

Shukovsky, Paul. 2007. Airport Profilers: They're Watching Your Expressions. *Seattle Post-Intelligencer*.
http://wwwseatlepi.nwsource.com/local/344868_airportprofiler26. html.

Silva, Denise Ferreira da. 2007. *Toward a Gobal Idea of Race*. Minneapolis and London: University of Minnesota Press.

Smallwood, Gracelyn. 2009. The Impact of Human Rights Violations on Aboriginal Health. Paper presented at the Human Rights Law and Practice Conference, 20–21 October, Melbourne, Australia.

Sniffen, Michael J. 2003. Pentagon anti-terror surveillance system hopes to identify people by the way they walk. The Associated Press, 19 May, http://www.securityfocus.com/news/4909.

Snyder, Sharon L. and David T. Mitchell. 2005. *Cultural Locations of Disability.* Chicago: University of Chicago Press.

Socolinsky, Diego A. 2008. Multispectral Face Recognition. In *Handbook of Biometrics*, ed. Anil K. Jain, Patrick Flynn and Arun A. Ross, 393–313. New York: Springer.

Somatechnics Research Centre. http://www.somatechnics.org.

Spivak, Gayatri Chakravorty. 1987. *In Other Worlds.* New York: Methuen.

Stacey, Jackie. 1999. Desperately Seeking Difference. In *Visual Culture: The Reader*, ed. Jessica Evans and Stuart Hall, 390–401. London: Sage.

Stafford, Barbara Maria. 1997. *Body Criticism: Imaging the Unseen in Enlightenment Art and Medicine.* Cambridge, MA: MIT Press.

Stryker, Susan. 1994. My Words to Victor Frankenstein Above the Village of Chamounix: Performing Transgender Rage. *GLQ: A Journal of Lesbian and Gay Studies* 1: 227–54.

——2006. (De)Subjugated Knowledges: An Introduction to Transgender Studies. In *The Transgender Studies Reader*, ed. Susan Stryker and Stephen Whittle, 1–17. New York and London: Routledge.

Tagg, John. 1988. *The Burden of Representation: Essays on Photographies and Histories.* Houndmills: Macmillan Press.

Talbot, Margaret. 2007. Duped: Can Brain Scans Uncover Lies? *The New Yorker*, 2 July, 52–61.

Tancredi, Laurence R. 2004. Neuroscience Developments and the Law. In *Neuroscience and the Law*, ed. Brent Garland, 71–113. New York: Dana Press.

Tanner, Dennis and Matthew Tanner. 2004. *Forensic Aspects of Speech Patterns: Voice Prints, Speaker Profiling, Lie and Intoxication Detection.* Tucson: Lawyers and Judges Publishing Company.

Tate, Greg, ed. 2003. *Everything But the Burden.* New York: Harlem Moon Broadway Books.

Tavernise, Sabrina. 2006. At least 600,000 civilians killed in Iraq, study finds. *Sydney Morning Herald*, October 12, 11.

Taylor, Paul. C. 2008. *Race: A Philosophical Introduction.* Cambridge: Polity.

Thacker, Eugene. 2003. Data Made Flesh: Biotechnology and the Discourse of the Posthuman. *Cultural Critique* 53: 72–97.

Thalheim, Lisa, Jan Krissler and Peter-Michael Ziegler. 2002. Body Check: Biometric Access Protection Devices and their Programs Put to the Test, *Technology Review*, c't 11/2002. http://www.heise.de/ct/english/02/11/114/.

Thandeka. 2000. *Learning to Be White.* New York and London: Continuum.

Tovino, Stacey A. 2007. Imaging Body Structure and Mapping Brain Function: A Historical Approach. *American Journal of Law and Medicine* 33: 193–228.

U.S. General Accounting Office. 2002. Technology Assessment: Using Biometrics for Border Security. http://www/goa.gov/htext/do3174.html.

UNHCR. 2003. Afghanistan: Iris-Testing Proves Successful. 10 October. http://www.unhcr.org/news/NEWS/3f86a3ac1.html.

Uttal, William R. 2001. *The New Phrenology: The Limits of Localizing Cognitive Processes in the Brain.* Cambridge, MA: MIT Press.

——2009. *Neuroscience in the Courtroom.* Tucson: Lawyers and Judges Publishing Company.

Valencia, Valorie S. with Christopher Horn. 2003. Biometric Liveness Testing. In *Biometrics: Identity Assurance in the Information Age*, ed. Woodward, John D., Jr., Nicholas M. Orlans and Peter T. Higgins, 139–49. Berkeley: McGraw-Hill/Osborne.

Van der Ploeg, Irma. 1999. The Illegal Body: "Eurodac" and the Politics of Biometric Identification. *Ethics and Information Technology* 1: 295–302.

——2005. *The Machine-Readable Body: Essays on Biometrics and the Informatization of the Body*. Maastricht: Shaker Publishing.

Van Dijck, José. 2005. *The Transparent Body: A Cultural Analysis of Medical Imaging*. Seattle and London: University of Washington Press.

Van Wyhe, John. 2004. *Phrenology and the Origins of Victorian Scientific Naturalism*. Aldershot: Ashgate.

Vanstone, Amanda. 2005. Border Security Enhanced by Biometrics, Passport Alert List Trials. http://www.minister.immi.gov.au/media_releases05/v05117.htm.

Welch, Dylan. 2008. Asylum-Seekers to go on Criminal Database. *Sydney Morning Herald*, 9 November. http://www.smh.com.au/news/national/asylumseekers-to-go-on-criminal-database/2008/11/08/1225561202058.html.

Williamson, Maurice. 2009. SmartGate Streamlines Trans-Tasman Travel. Beehive.govt.nz: The official website of the New Zealand Government. http://beehive.govt.nz/release/smartgate+stremlines+trans-tasman+travel.

Wilson, Dean. 2007. Australian Biometrics and Global Surveillance. *International Criminal Justice Review* 17: 207–19.

Wolpe, Paul Root, Kenneth R. Foster and Daniel D. Langleben. 2005. Emerging Neurotechnologies for Lie-Detection: Promises and Perils. *American Journal of Bioethics* 5: 39–49.

——2005a. Response to Commentators on "Emerging Neurotechnologies for Lie-Detection: Promises and Perils?" *American Journal of Bioethics* 5: W5.

Woodward Jr., John D., Katherine W. Webb, Elaine M. Newton, Melissa Bradley and David Rubenson. 2001. *Army Biometric Applications*. Santa Monica: RAND.

Woodward Jr., John D., Nicholas M. Orlans and Peter T. Higgins. 2003. *Biometrics: Identity Assurance in the Information Age*. Berkeley: McGraw-Hill/Osborne.

Woodward, Jr., John D. 2005. Using Biometrics to Achieve Identity Dominance in the Global War on Terrorism. *Military Review* (September-October): 30–34.

——2004. How Do You Know Friend from Foe? *Homeland Science and Technology*, December: 112–13.

Wright, Jim. 2000. Brain Dynamics: Modelling the Whole Brain in Action. In *Integrative Neuroscience*, ed. Evian Gordon, 139–71. Amsterdam: Harwood Academic Publishers.

Yancy, George, ed. 2005. *White on White/Black on Black*. Lanham: Rowman and Littlefield.

Yoo, Jang-Hee, Mark S. Nixon and Chris J. Harris. 2002. Extracting Gait Signatures Based on Anatomical Knowledge. http://www.bmva.ac.uk/meetings/meetings/02/6March02/soton2.pdf.

Index

Kolibri 360 GmbH
Ferdinand-Jühlke-Straße 7
99095 Erfurt, Deutschland
produktsicherheit@kolibri360.de